AutoCAD Civil 3D 2024 for Novices

(Learn By Doing)

CADSoft Technologies

CADSoft Technologies

AutoCAD Civil 3D 2024 for Novices:
Learn By Doing

CADSoft Technologies

ISBN: 978-1-942689-62-1

DEDICATION

To students, who are dedicated to learning new technologies

THANKS

To employees of CADSoft Technologies for their cooperation

Online Training Program Offered by CADSoft Technologies

CADSoft Technologies provides effective and affordable virtual online training on various software packages including Computer Aided Design, Manufacturing and Engineering (CAD/CAM/CAE), computer programming languages, animation, architecture, and GIS. The training is delivered 'live' via Internet at any time, any place, and at any pace to individuals as well as the students of colleges, universities, and CAD/CAM/CAE training centers. The main features of this program are:

Training for Students and Companies in a Classroom Setting

Highly experienced instructors and qualified Engineers at CADSoft Technologies conduct the classes under the guidance of Prof. Sham Tickoo of Purdue University Northwest, USA. This team has authored several textbooks that are rated "one of the best" in their categories and are used in various colleges, universities, and training centers in North America, Europe, and in other parts of the world.

Training for Individuals

CADSoft Technologies with its cost effective and time saving initiative strives to deliver the training in the comfort of your home or work place, thereby relieving you from the hassles of traveling to training centers.

Training Offered on Software Packages

CADSoft Technologies provides basic and advanced training on the following software packages:

CAD/CAM/CAE: *CATIA, Creo Parameteric, SOLIDWORS, Autodesk Inventor, Solid Edge, NX, AutoCAD, AutoCAD LT, Customizing AutoCAD, AutoCAD Electrical, and ANSYS*

Architecture and GIS: *Autodesk Revit (Architecture/Structure/MEP), AutoCAD Civil 3D, AutoCAD Map 3D, Autodesk Navisworks, Bentley Staad.Pro, Oracle Primavera P6, MS Project, ArcGIS, and Raster Design.*

Animation and Styling: *Autodesk 3ds Max, 3ds Max Design, Autodesk Maya, Autodesk Alias, Pixologic ZBrush, and CINEMA 4D*

Computer Programming: *C++, VB.NET, Oracle, AJAX, and Java*

For more information, please visit the following link:
https://cadsofttech.com

Table of Contents

Chapter 2: Working with Points

Chapter 3: Working with Surfaces

Chapter 4: Surface Volumes and Analysis

Chapter 5: Alignments

Chapter 6: Working with Profiles

Chapter 7: Working with Assemblies and Subassemblies

Chapter 8: Working with Corridors and Parcels

Chapter 9: Sample Lines, Sections, and Quantity Takeoffs

Chapter 10: Feature Lines and Grading

Chapter 11: Pipe Networks

Chapter 12: Pressure Networks

Chapter 13: Working with Plan Production Tools and Data Shortcuts

This page is intentionally left blank

Preface

AutoCAD Civil 3D 2024

AutoCAD Civil 3D is one of the leading civil engineering software developed by Autodesk Inc., USA. This software provides a comprehensive solution for various infrastructure works such as earthwork, subdivision, highways, land development, survey, drainage, and different areas related to project management. AutoCAD Civil 3D has the ability to dynamically link various objects created in it. As a result any changes or revisions made in the design are instantly updated by a click of mouse. This powerful feature saves a lot of time, thereby enabling the engineers and designers involved in the project to assess various conditions that can affect their design process.

AutoCAD Civil 3D 2024 for Novices textbook has been written with the intention of helping the readers effectively use AutoCAD Civil 3D. Structured in a pedagogical sequence, the chapters in this textbook cover the basic as well as advanced concepts of AutoCAD Civil 3D such as alignments, surfaces, gradings, corridor modeling, and earthwork calculations to name only a few. The civil engineering industry examples that are used as tutorials and the related additional exercises at the end of each chapter will help the users to understand the techniques used in the industry to design a project.

The salient features of this textbook are as follows:

- **Tutorial Approach**
  The author has adopted the tutorial point-of-view and the learn-by-doing approach throughout the textbook. This approach guides the users through the process of creating models in the tutorials.

- **Real-World Models as Projects**
  The author has used about 36 real-world projects as tutorials in this textbook. This enables the readers to relate the models in the tutorials to the real-world. In addition, there are about 20 exercises that can be used by the readers to assess their knowledge.

- **Tips and Notes**
  Additional information related to various topics is provided to the users in the form of tips and notes.

- **Heavily Illustrated Text**
  The text in this textbook is heavily illustrated with about 750 diagrams and screen capture images.

- **Learning Objectives**
  The first page of every chapter summarizes the topics that are covered in the chapter.

- **Exercises**
  The Exercises are given at the end of the chapters and they can be used by the instructor as test questions and assignments.

Symbols Used in the Textbook

Note

The author has provided additional information to the users about the topic being discussed in the form of notes.

Tip

Special information and techniques are provided in the form of tips that help in increasing the efficiency of the users.

Formatting Conventions Used in the Textbook

Please refer to the following list for the formatting conventions used in this textbook.

- Names of tools, buttons, options, browser, palette, panels, and tabs are written in boldface.
  Example: The **Create Surface** tool, the **Modify** button, the **Create Design** panel, the **Home** tab, **Properties Palette**, **Content Browser**, and so on.

- Names of dialog boxes, drop-downs, drop-down lists, list boxes, areas, edit boxes, check boxes, and radio buttons are written in boldface.
  Example: The **Table Style** dialog box, the **Points** drop-down of **Create Ground Data** panel in the **Home** tab, the **Name** edit box of the **Surface Properties** dialog box, and so on.

- Values entered in edit boxes are written in boldface.
  Example: Enter **EG Surface** in the **Name** edit box.

- Names of the files saved are italicized.
  Example: *c03_tut1a.dwg*

- The methods of invoking a tool/option from the ribbon, Application Menu, or the shortcut keys are given in a shaded box.

Ribbon:	Home > Create Ground Data > Surface drop-down > Create Surface
Command:	CREATESURFACE

Naming Conventions Used in the Textbook

Tool

If you click on an item in a panel of the ribbon and a command is invoked to create/edit an object or perform some action then that item is termed as **tool**.

For example:
Create Corridor tool, **Create Surface** tool, **Create Surface Profile** tool

If you click on an item in a panel of the ribbon and a dialog box is invoked wherein you can set the properties to create/edit an object then that item is also termed as **tool**, refer to Figure 1.

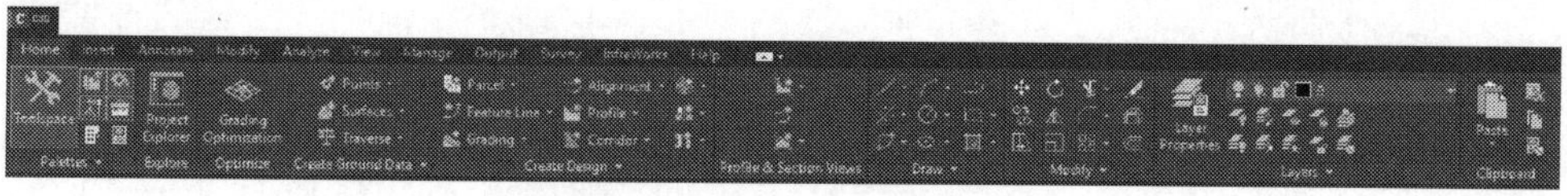

Figure 1 *Tools in the Ribbon*

For example:
Create Surface tool, **Create Feature Line** tool
Water Drop tool, **Bounded Volumes** tool

Dialog Box

In this textbook, different terms are used for referring to the components of a dialog box. Refer to Figure 2 for the terminology used.

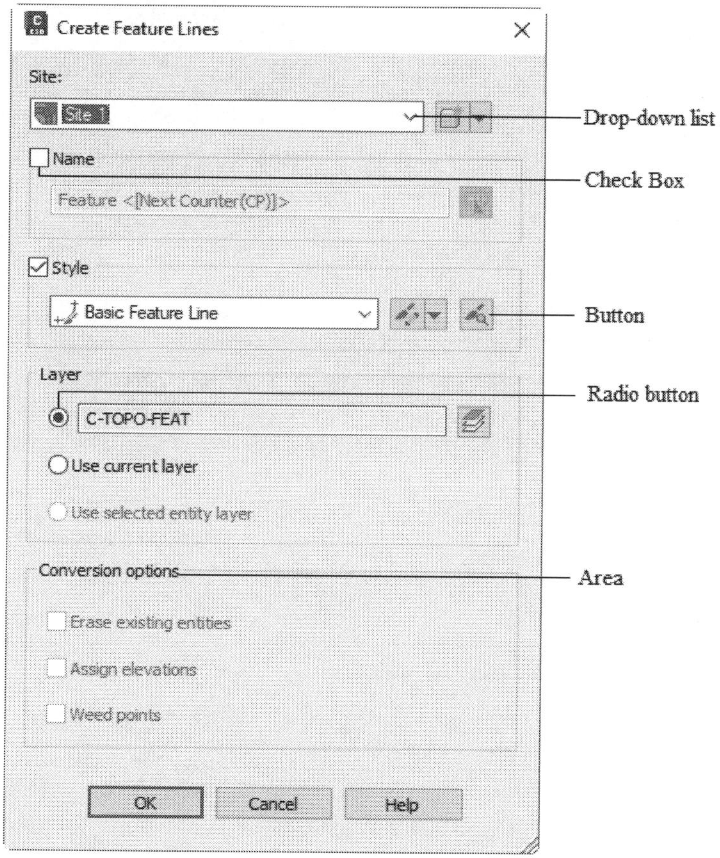

Figure 2 *Different terminologies used in a dialog box*

Button

The item in a dialog box that has a 3D shape like a button is termed as **button**. For example, **OK** button, **Cancel** button, **Apply** button, and so on. If the item in a ribbon is used to exit a tool or a mode, it is also termed as button. For example, **Modify** button, **Finish Editing System** button, **Cancel Editing System** button, and so on, refer to Figure 2.

Drop-down

A drop-down is one in which a set of common tools are grouped together. You can identify a drop-down with a down arrow on it. These drop-downs are given a name based on the tools grouped in them. For example, **Surfaces** drop-down, **Points** drop-down, **Grading** drop-down, and so on, refer to Figure 3.

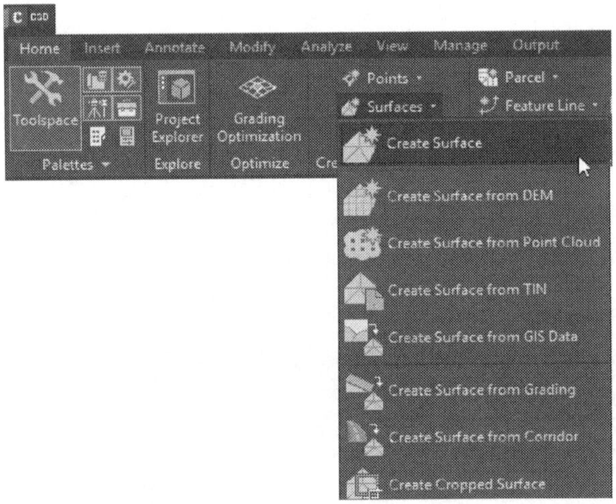

Figure 3 *Choosing a tool from drop-down*

Drop-down List

A drop-down list is one in which a set of options are grouped together. You can set various parameters using these options. You can identify a drop-down list with a down arrow on it. For example, **Type Selector** drop-down list, **Units** drop-down list, and so on, refer to Figure 4.

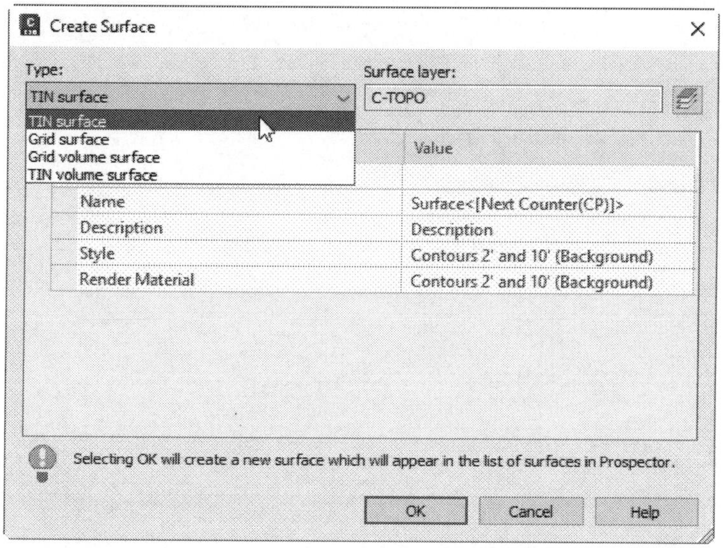

***Figure 4** Selecting an option from the **Type** drop-down list*

Options

Options are the items that are available in shortcut menu, drop-down list, dialog boxes, and so on. For example, choose the **Alignment Properties** option from the shortcut menu displayed on right-clicking in the drawing area, refer to Figure 5.

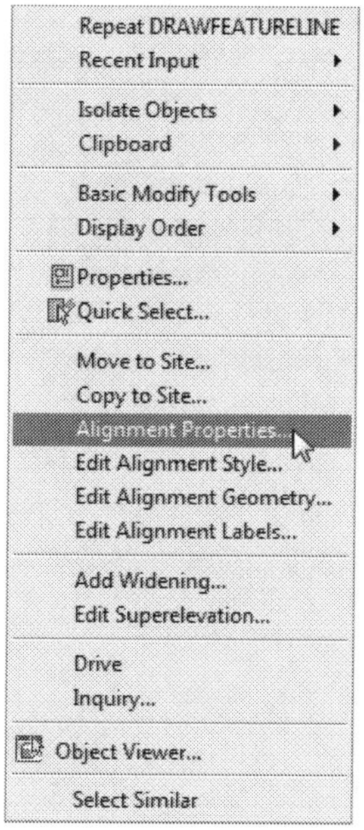

***Figure 5** Choosing an option from the shortcut menu*

This page is intentionally left blank

Chapter 1

Introduction to AutoCAD Civil 3D 2024

Learning Objectives

After completing this chapter, you will be able to:

- *Understand the basic features of AutoCAD Civil 3D 2024*
- *Understand AutoCAD Civil 3D 2024 screen components*
- *Get familiar with different workspaces in AutoCAD Civil 3D 2024*

INTRODUCTION TO AutoCAD Civil 3D 2024

AutoCAD Civil 3D is a powerful Building Information Modeling (BIM) tool that is used for designing, analyzing, and documenting engineering projects such as land development, transportation, and environmental projects.

This software includes various sets of tools for designing a civil engineering project along with the tools from other Autodesk products such as AutoCAD and Map 3D. It also has interfacing application such as Trimble Link™, Carlson Connect™, Infraworks, and Leica Exchange for transferring and converting data set into Civil 3D environment. These tools can be accessed through toolbars and Ribbon provided in the software interface. Civil 3D also supports the traditional commands that exist in AutoCAD software to invoke various tools. These commands are executed using the command prompt in the software.

AutoCAD Civil 3D 2024 provides a model-centric technology at the core that enables the entire team in a project to coordinate and work on a single updated model. Data sharing is conducted through various approaches ranging from Xrefs and data shortcuts to integrated data management controls. These approaches help team members to work in parallel, connected locally or remotely, on the latest updated model. As a result, a project can be managed well and the budget can also be controlled.

BASIC FEATURES OF AutoCAD Civil 3D

AutoCAD Civil 3D features a range of design and analysis objects within its three-dimensional dynamic engineering model. These Civil 3D objects include points, surfaces, parcels, alignments, profiles, and gradings. All objects in this model have a hierarchical interaction mechanism with each other. This mechanism ensures that modifications carried out in one object are truly reflected in other objects that are connected to it.

A Civil 3D object is a drawing element that maintains a relationship with another Civil 3D objects. For example, an alignment object is a combination of lines and curves in a horizontal plane that collectively defines the location of a project component such as the centerline of a road. This alignment can be an independent object or it can be a parent object of other Civil 3D objects such as profiles and cross-sections. If you edit this alignment object it will result in the change of other related Civil 3D objects.

Civil 3D has other objects such as points, surfaces, alignments, profiles, sites, parcels, gradings, corridors, assemblies, and subassemblies. All these objects have a designated hierarchy that can be viewed in an interface component called the **TOOLSPACE** palette.

AutoCAD Civil 3D also features various other object components such as tables, object labels, and the analysis results that are derived from the model. These object components are associated with the core model and they can be dynamically updated so as to reflect any change made in the core model.

Some of the various objects required to create a Civil 3D model are discussed next.

Points
Points are the basic building blocks in AutoCAD Civil 3D. In Civil 3D, points are coordinate geometry points and are also known as the COGO points. In a civil engineering project, points are used to represent the existing ground locations and design elements. A Civil 3D point object is different from the AutoCAD point object. An AutoCAD point merely displays the location of an object. A Civil 3D point has a unique identification number and properties such as northing, easting, elevation, and description. These points can also have additional properties to control their appearance such as point style, point label style, and layer.

Point Groups
The point groups allow you to group the COGO points together. The groups are created by using the point objects that have similar characteristics. This helps you to control the overall appearance of the points. Point group objects make it easy to change the point number, point style, and other properties of a large number of points.

Surfaces
Surfaces are key objects in AutoCAD Civil 3D. They are the three-dimensional geometric representation of a land terrain. In Civil 3D, you can calculate a composite volume surface that represents the difference between two surface areas.

Alignments
The alignment objects in civil engineering projects represent the geometry of features like the pipeline, road, canal, railway, or construction baselines in a horizontal plane. Creating and defining the alignment is the first step in the design process of highway, railway, or other engineering projects.

Profiles
The profiles (or long sections) are used to show surface levels along a selected horizontal alignment using the specified surfaces. After creating a profile in Civil 3D, you can design the vertical alignment directly on it by using standard alignment tools, grips, or editors.

Profile Views
The profile views are used to represent the graph lines which form the grid along which the profile is plotted. One profile view object can be used to represent multiple profiles. Profile view object also includes data bands. Data bands can be used to display additional information about the profile view.

Assemblies and Subassemblies
Assemblies constitute the primary structure of an AutoCAD Civil 3D corridor model. The subassemblies are the building blocks for creating an assembly. The assembly objects are composed of subassemblies. The assembly forms the primary structure of a corridor model. The subassemblies can be carriage ways, curbs, and slopes that can be added to the assembly baseline to create an assembly.

Corridors

A corridor represents a path such as a road, railway, or canal. The horizontal and vertical geometry of a corridor is defined by the alignment and the profile.

The corridor objects are created along one or more baselines (alignments). These are then created by placing a 2D section (assembly) at incremental locations along the baselines and by creating matching slopes that reach a surface model at each incremental location. The corridor models can be edited on the basis of sections.

Parcels

The parcels represent any closed boundary that is spatially referenced to a geographical location. However, the major use of parcel is to represent a tract of land which could be a lot in a subdivision. Parcel can also be a water body or even a lot in a given site which is termed as site parcel.

Grading

In AutoCAD Civil 3D, the grading tools enable the engineers to grade a surface by applying different grading criteria. On applying grading to a surface, you can analyze the grading and balance the cut and fill volumes of a surface.

Sections

In AutoCAD Civil 3D, sections or cross-sections provide a view of the terrain that is cut at an angle across a linear feature such as a proposed road. The sections are cut across the horizontal alignments at specified station intervals. These sections can be plotted individually for a specified station or as a group for a specified range of stations, depending on the purpose of a plot. You can generate the section volumes from a design for earthworks and materials using the **Quantity Take Off** tools.

Pipe Networks

You can use pipe networks to design the utility system such as the storm and wastewater system which carry fluids under the effect of gravity. The pipe networks are created from design catalogs and can be edited in plan views. You can also display the pipe network parts in profile and section views. Therefore, if a change is made to the pipe networks in plan view, the profile and section views are dynamically updated.

Pressure Pipe Networks

In Civil 3D, the pressure pipe network consists of objects that are used to design a pipe network that carries fluids under pressure, such as water supply network. In this category, various tools are provided to design the network in plan and profile layout. Similar to pipe network, the pressure pipe network is created from the parts that are available in a parts list. This parts list is derived from the AutoCAD catalog which is a master collection of various available parts.

Superelevation View

Superelevation or banking is defined as the transverse inclination to the road pavement surface created to counter the centrifugal forces that are exerted on the vehicle while travelling along a curved path. The superelevation view in Civil 3D is used to represent the superelevation. It is also used to perform the superelevation calculations and edit the superelevation data.

STARTING AutoCAD Civil 3D 2024

You can start AutoCAD Civil 3D 2024 by double-clicking on its shortcut icon on the desktop or from the taskbar. To start it from the taskbar, click on Windows icon and then choose **Civil 3D 2024 Imperial/Metric** from the menu displayed (for Windows 10), refer to Figure 1-1; AutoCAD Civil 3D 2024 will start and a window will be displayed with the **Start Tab** chosen by default.

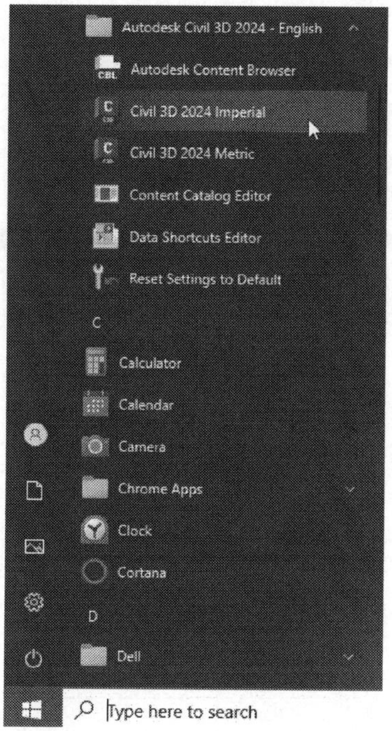

Figure 1-1 *Starting AutoCAD Civil 3D 2024 from the taskbar*

AutoCAD Civil 3D 2024 USER INTERFACE

The Civil 3D interface consists of various components such as the drawing area, command window, toobars, Ribbon, model and layout tabs, Status Bar, **TOOSPACE** palette, and **PANORAMA** window, as shown in Figure 1-2. The title bar is displayed on the top of the screen and displays the current drawing name.

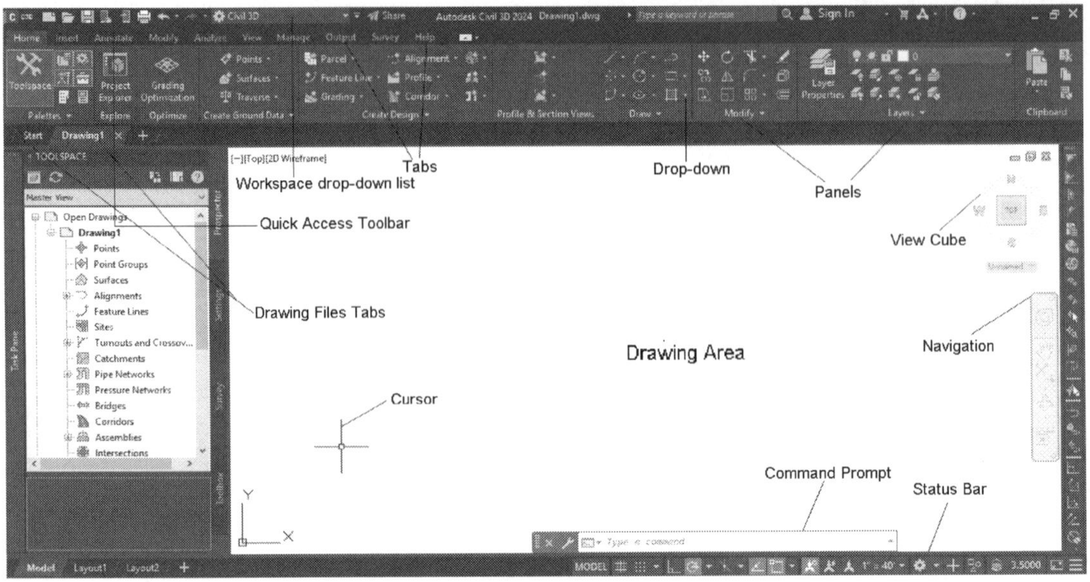

***Figure 1-2** AutoCAD Civil 3D 2024 interface components*

The AutoCAD Civil 3D interface components are discussed next.

Drawing Area

The drawing area covers the major portion of the screen. Here, you can draw various objects and use commands. To draw the objects, you need to define the coordinate points that can be selected by using a pointing device. The position of the pointing device is represented on the screen by the cursor. There is a coordinate system icon at the lower left corner of the drawing area. The window also has standard Windows buttons such as close, minimize, scroll bar, and so on at the top right corner. These buttons have the same functions as in any other standard window.

Ribbon

The Ribbon interface in AutoCAD Civil 3D has a collection of tools that are grouped into various tabs based on their functionality. This interface provides an alternative to the layered menus, toolbars, and task panes which were used previously for displaying tools. The tools in the tabs are further sub grouped and placed into various panels depending on their usage and functionality. To invoke a command, choose the required tool from the Ribbon. Figure 1-3 shows the typical **Home** tab of the AutoCAD Civil 3D Ribbon.

***Figure 1-3** Complete view of the Ribbon showing the **Home** tab*

Application Menu

The **Application Menu**, shown in Figure 1-4, will be displayed when you choose the **Application** button located on the top left corner of the Civil 3D window.

It contains some of the tools that are available in the **Standard** toolbar. You can search a command using the search field in the **Application Menu**. To search a tool, enter the complete or partial name of the command in the search field; the possible tool list will be displayed. If you click on a tool from the list, the corresponding command will get activated.

By default, the **Recent Documents** button is chosen in the **Application Menu**. As a result, the list of recently opened drawings will be displayed in the menu. If you have opened multiple drawing files, choose the **Open Documents** button; the documents that are opened will be listed in the **Application Menu**. To set the preferences of the file, choose the **Options** button available at the bottom right of the **Application Menu**. To exit Civil3D, choose the **Exit Autodesk Civil 3D 2024** button next to the **Options** button.

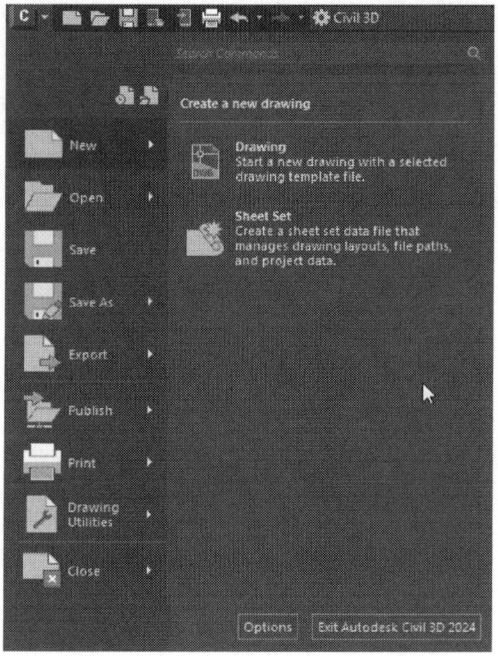

Figure 1-4** The **Application Menu

Command Bar

The command bar located at the bottom of the drawing area has the command prompt where you can enter the commands. It also displays the subsequent prompt sequences and messages, refer to Figure 1-2. You can change the size of the command bar by placing the cursor on the top edge (double line bar known as the Grab bar) and then dragging it. This way you can increase its size to see the previous commands that you have used. By default, the command window displays only three lines. You can also press the F2 function key to display the **AutoCAD Text** window.

Drawing File Tabs

The drawing file tabs are displayed above the drawing area, refer to Figure 1-2, shows drawing that are currently opened. Using these drawing tabs, you can quickly switch between the drawings. The order in which these tabs are displayed is based on the sequence in which the files are opened.

Status Bar

The Status bar is displayed at the bottom of the screen, as shown in Figure 1-5. It contains some useful information and buttons that will make it easier for you to change the status of some AutoCAD functions. You can toggle between on and off states of most of these functions by choosing buttons from this bar. The button are:

Model
Display Drawing Grids
Snap Mode
Ortho Mode
Polar Tracking
Isometric Drafting
Object Snap Tracking
Object Snap
Show /Hide Lineweight
Show Annotation Objects
Add Scales to Annotative Objects when the Annotative Scale Changes
Annotation Monitor
Isolate Objects
Hardware Acceleration On
Clean Screen
Customization

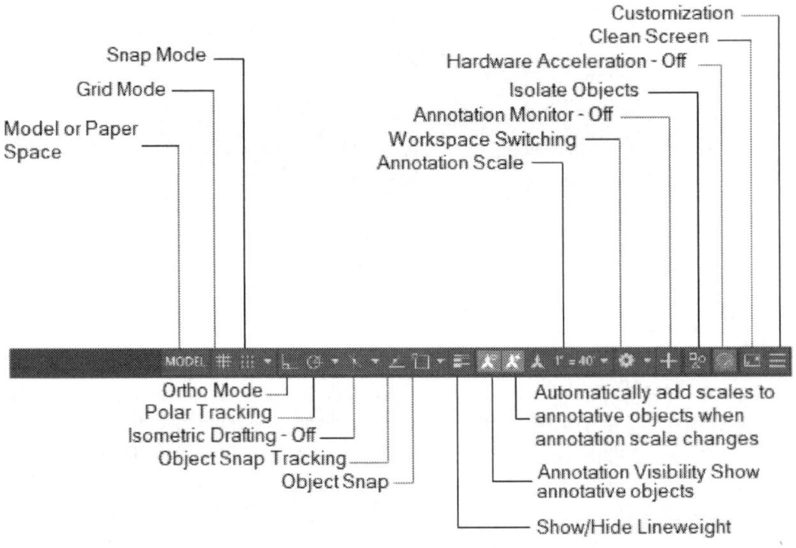

Figure 1-5 *The Status bar displayed in the Civil 3D workspace*

TOOLSPACE Palette

In AutoCAD Civil 3D, you can view the project data and its status at any time in the **TOOLSPACE** palette. The data in the palette updates itself dynamically to show the status of the data within the drawing. To display the **TOOLSPACE** palette in a drawing, choose the **TOOLSPACE** button from the **Palettes** panel of the **Home** tab or type **SHOWTS** in the command line. There are four tabs in the palette:

Prospector Tab
Settings Tab
Survey Tab
Toolbox Tab

Shortcut Menu

AutoCAD Civil 3D provides you with shortcut menus that are an easy and convenient way of invoking tools and the recently used commands. These shortcut menus are context-sensitive which means that the commands present in them are dependent on the place or object for which they are displayed. A shortcut menu is invoked by right-clicking and is displayed at the cursor location. You can right-click anywhere in the drawing area to display the general shortcut menu. The shortcut menus are also displayed when you right-click on one or more drawing objects or on an individual item from the **TOOLSPACE** palette. Figure 1-6 shows a shortcut menu displayed on right-clicking on the **Points** node in the **Prospector** tab.

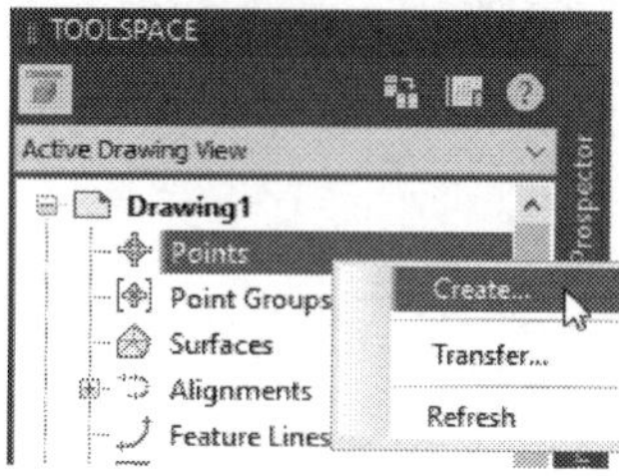

***Figure 1-6** Shortcut menu displayed on right-clicking on the **Points** node in the **Prospector** tab*

Layout Tools

In AutoCAD Civil 3D, you can use various layout toolbars to create and edit the Civil 3D objects such as alignments, grading, points, profiles, pipes, pressure pipes, and parcels. These toolbars provide access to object-specific designs and edit commands. To invoke a toolbar, choose the drop-down corresponding to the object. Next, choose the layout or creation method from the options that will be displayed in the flyout. Figure 1-7 shows the **Alignment Layout Tools - <alignment name>** toolbar invoked by choosing **Alignment Creation Tools** from **Home > Create Design > Alignment** drop-down.

***Figure 1-7** The **Alignment Layout Tools - <alignment name>** toolbar*

Autodesk Account

You can access the Autodesk services using your Autodesk ID. To login to your Autodesk account, choose the **Sign In** button in the **InfoCenter** toolbar; a drop-down list will be displayed. Next, choose the **Sign In to Autodesk account** option from this list; the **Autodesk-Sign In** dialog box will be displayed, as shown in Figure 1-8. Enter your credentials in this dialog box and choose the **Sign In** button; Autodesk will validate your credentials and then provide access to your account.

Tip
You can sign up for a Autodesk ID by choosing the ***CREATE ACCOUNT*** *button in the* ***Autodesk - Sign In*** *dialog box.*

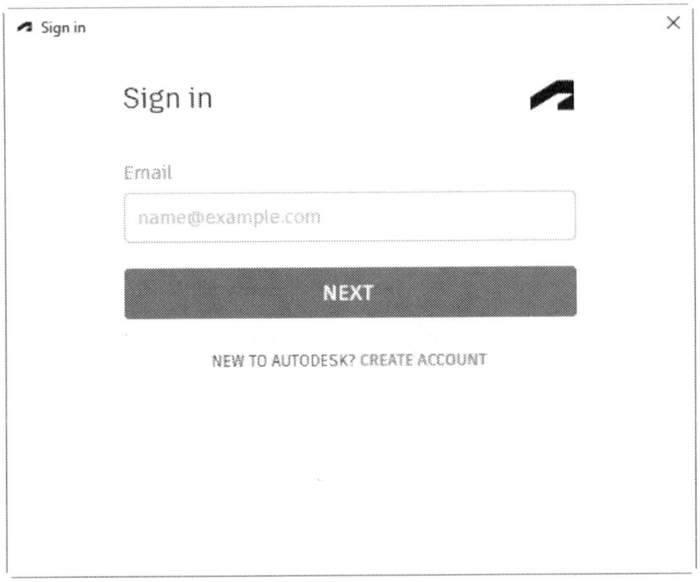

Figure 1-8 The ***Autodesk - Sign In*** *dialog box*

PANORAMA Window

This window can be used to display the entities of the objects such as alignments and profiles in a tabular form. The **PANORAMA** window is a floating, dockable window that can be kept open while working. This window includes several tables called vistas on different tabs. In the **PANORAMA** window, the data displayed in black is editable and the data displayed in grey is non-editable.

Civil 3D WORKSPACES

Workspaces are the set of toolbars, menus, and Ribbon that are grouped and organized together to work in a custom and task-oriented environment. A workspace controls the layout of the Civil 3D user interface. When you use a specific workspace, you can view the Ribbon, toolbars, and other applications that have been specified for the selected workspace. To invoke the applications and tools that are specified in other workspaces, enter the commands in the command line.

You can select workspaces at the beginning when you start the Civil 3D application or anytime during the drawing session. When you start AutoCAD Civil 3D, you are prompted to select a specific workspace and then continue. Once you are done, you can switch to other workspace settings at any time by choosing the **Workspace Switching** button that is located in the Status Bar. On doing so, a flyout will be displayed, as shown in Figure 1-10.

Choose an option from this flyout for specifying different workspace settings in the current drawing environments. After you choose the required option, the drawing interface will change according to the setting of the workspace you have chosen.

In AutoCAD Civil 3D, there are four default workspaces such as **Civil 3D**, **Planning and Analysis**, **2D Drafting & Annotation**, and **3D Modeling**. These default workspaces are designed considering the tools that a user might use for that kind of project. You can also create your own customized workspace and save it using the **Save Current As** tool from the flyout, refer to Figure 1-9.

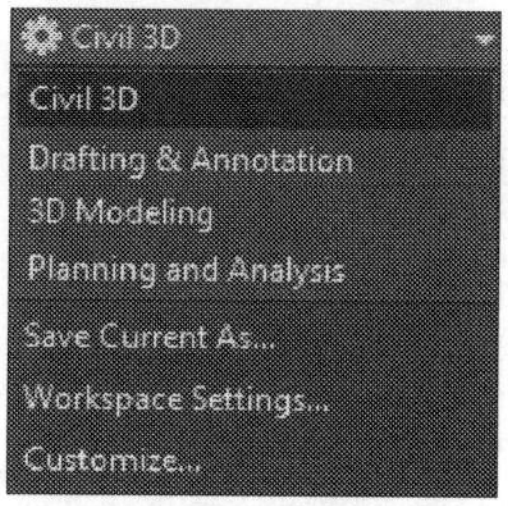

***Figure 1-9** The flyout displaying various options for choosing a workspace*

The default workspaces that are available in AutoCAD Civil 3D are discussed next.

Civil 3D

The Civil 3D workspace includes all the necessary tools and applications that required for designing a civil engineering project such as the road, railway, or a pipe network designing project.

Drafting & Annotation

This workspace is used for the purpose of documentation of a project. The **Annotation** tab is included in the Ribbon in this workspace. All the tools related to annotation are available in this tab so that you can annotate easily as per your requirement.

3D Modeling

This workspace contains all the tools and applications that are available in the AutoCAD and can be useful in AutoCAD Civil 3D.

Planning and Analysis

This workspace contains all the tools and features that are available in AutoCAD Map 3D and can be used directly in AutoCAD Civil 3D.

GETTING STARTED WITH AutoCAD Civil 3D

Before you create a new project or start working on the existing project, you should be familiar with various tools and concepts that can be applied to create the initial setup of the project. The basic setup tools like drawing templates and setting units are discussed next.

Drawing Templates

Drawing templates are very useful tools for AutoCAD Civil 3D users. These tools help in maintaining consistency throughout the drawings in the project.

Drawing Settings

Drawing settings allow the user to review or modify settings specified for a drawing. To edit the current drawing settings, right-click on the drawing name that is displayed in the **Settings** tab of the **TOOLSPACE** palette; a shortcut menu will be displayed. Choose the **Edit Drawing Settings** option from this shortcut menu, as shown in Figure 1-10; the **Drawing Settings - <drawing name>** dialog box will be displayed, refer to Figure 1-11. This dialog box contains five tabs that are:

Unit and Zone Tab
Transformation Tab
Object Layers Tab
Abbreviation Tab
Ambient Setting Tab

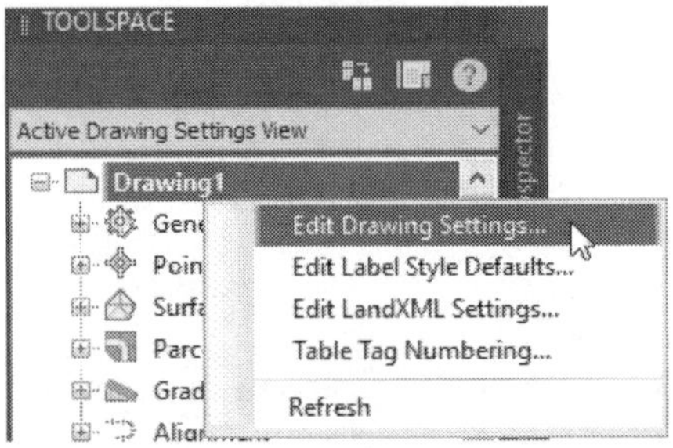

***Figure 1-10** Choosing the **Edit Drawing Settings** option*

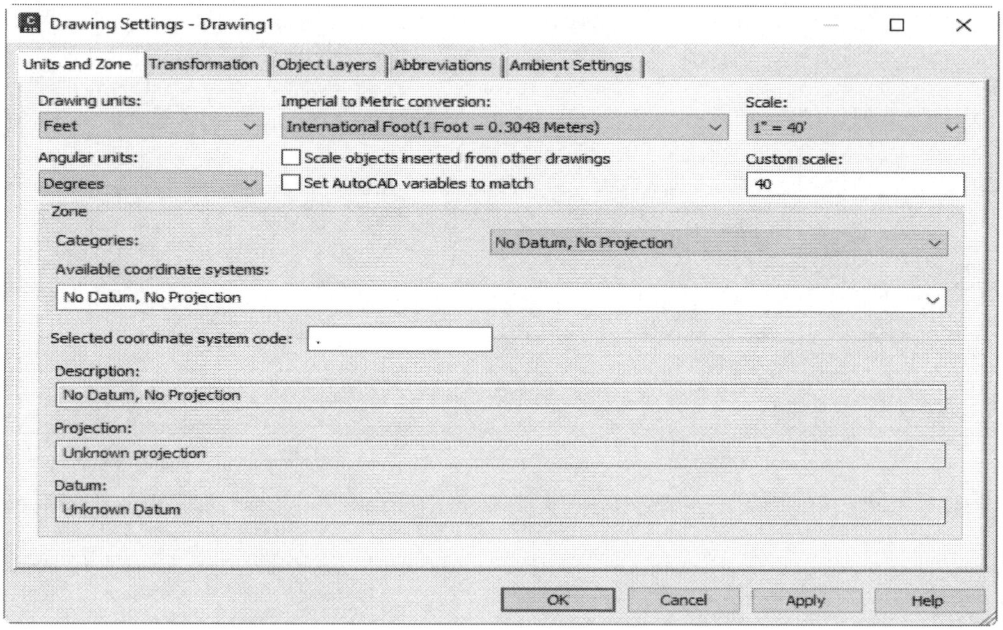

*Figure 1-11 The **Drawing Settings** dialog box*

Chapter 2

Working with Points

Learning Objectives

After completing this chapter, you will be able to:

- *Understand the concept of points in Civil 3D*
- *Create point styles*
- *Create point label styles*
- *Edit points*

POINT OBJECT

A point object represents a location in 3D space, and is defined by its X, Y, and Z coordinates. Each civil engineering project starts with data collection. This data is imported into Civil 3D workspace as points by using a suitable coordinate system. Generally, points specify the location of different features on the site such as trees, road geometry points, property corners, edge of pavements, and so on. However, Civil 3D points offer you more than just specifying the location of different objects on the site. In Civil 3D, each point represents an individual object with different information and has a unique point number. The information displayed by a Civil 3D point object depends upon the point settings, which will be discussed later in the chapter.

Components of Point Object

In AutoCAD Civil 3D, a point object has two major components, marker and label. Marker represents the location of point object whereas label displays information about that point object. The display of point marker is controlled by point style and the display of the point label is controlled by the point label style. Figure 2-1 shows a point with number, elevation, and description.

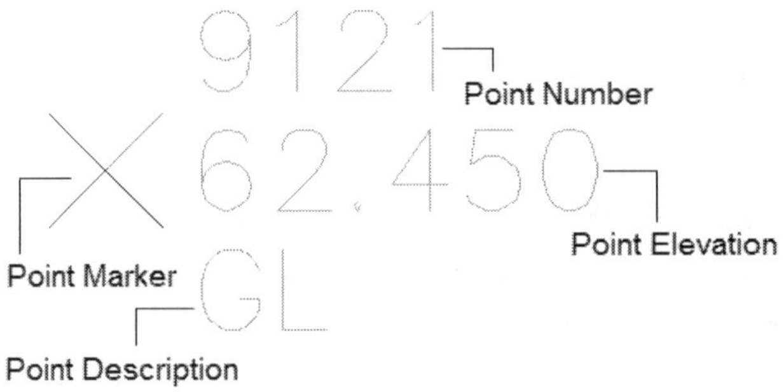

Figure 2-1 *A point with number, elevation, and description*

The properties of an individual Civil 3D point can be viewed in the **PROPERTIES** palette, refer to Figure 2-2. You can modify the properties of the selected point(s) such as color, layer, linetype, and point elevation.

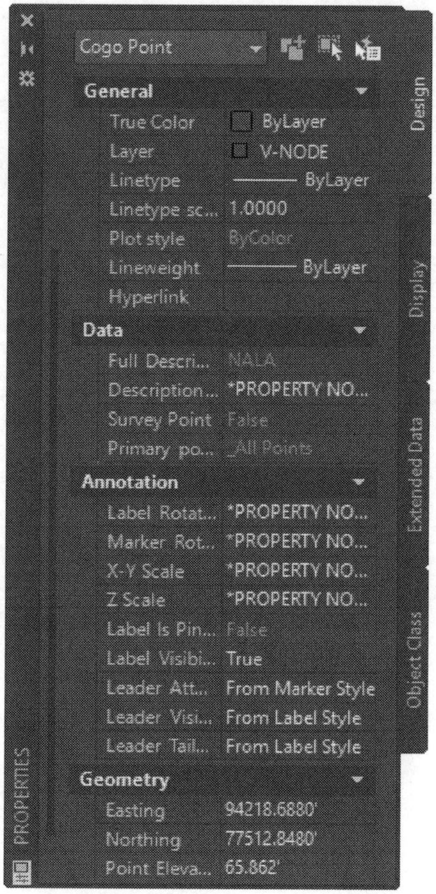

***Figure 2-2** Partial view of the **PROPERTIES** palette showing different point properties*

Creating Points

In AutoCAD Civil 3D, you can create points using various methods. These methods of point creation involve conversion of AutoCAD points into Civil 3D points, importing points from external point files, and creation of points using input parameters specified by the user.

POINT SETTINGS AND STYLES

AutoCAD Civil 3D has some in-built point styles that control the visibility and appearance of Civil 3D points. Besides controlling the visibility and display of points, these styles help you manage the workability of points. Point styles can be assigned after the points are imported or created. Alternatively, you can use a drawing template with the required point settings and then import or create points according to industry standards. Civil 3D has some default settings for points. These settings control display, elevation, layer, visibility, size, dimension, and so on. You can view and edit these default settings as per your requirement. To edit these settings, choose the **Settings** tab in the **TOOLSPACE** palette and right-click on the **Point** node; a shortcut menu will be displayed. Choose the **Edit Feature Settings** option from the shortcut menu, as shown in Figure 2-3; the **Edit Feature Settings - Point** dialog box will be displayed, as shown in Figure 2-4.

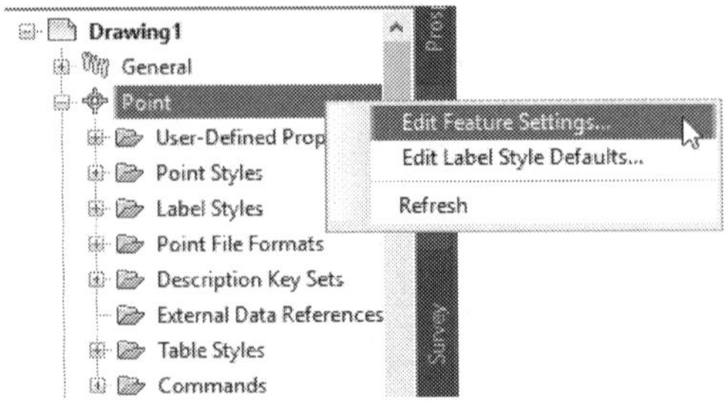

***Figure 2-3** Choosing the **Edit Feature Settings** option from the shortcut menu*

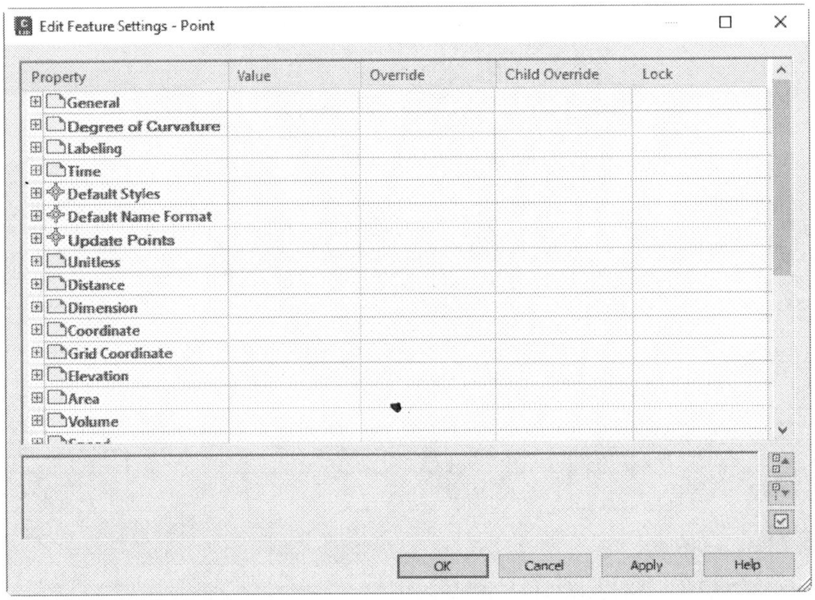

***Figure 2-4** The **Edit Feature Settings - Point** dialog box*

Point Styles

Point styles control the shape, size, color, and location of the point marker. It also controls the visibility of the point label. Point styles are created and assigned before or after points are created in the drawing. As discussed earlier, AutoCAD Civil 3D provides you with some in-built point styles. However, you can create your own point styles as per the project requirements and use them in the drawing. Point styles are created and managed in the **Settings** tab of the **TOOLSPACE** palette.

POINT LABEL STYLES

Point label styles control the behavior and appearance of point labels. Point labels provide information about points. Like point styles, point label styles are created and managed in the **Settings** tab of the **TOOLSPACE** palette.

EDITING POINTS

Ribbon:	Modify > Ground Data > Points
Command:	EDITPOINTS

You can edit the properties of a point. There are various tools that can be used to edit the properties of a point. To edit a property, choose the **Points** tool from the **Ground Data** panel of the **Modify** tab; the **COGO Point** contextual tab will be displayed in the Ribbon, as shown in Figure 2-5.

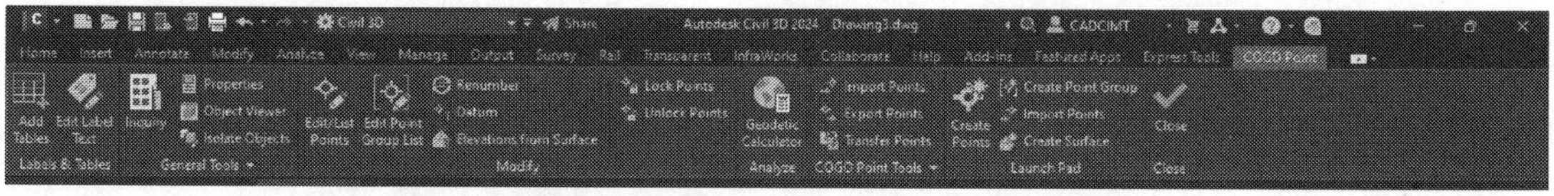

***Figure 2-5** The **COGO Point** contextual tab*

POINT GROUPS

As mentioned earlier, a point group is a collection of similar points. Point groups help you control, organize, and manage a point easily. Like points, point groups also have their own styles and label styles. You can assign different point styles and point label styles to each point group and control the appearance of points of a point group. It helps you quickly identify and modify point styles or point label styles of the required points of any point group. For example, you can group surface points that are used for creating surface in a group and assign a point style to them. Also, you can create another point group containing points of the centerline of an alignment and assign them a different point style and label style. Now, you can easily identify and distinguish the surface creation points from the centerline points in the drawing. These points can then be easily edited or managed in the drawing as required.

Creating Point Groups

Ribbon:	Home > Create Ground Data > Points drop-down > Create Point Group
Command:	CREATEPOINTGROUP

Point groups are created based on point properties. You can group points based on their point numbers, elevation, description, and so on. To create a point group, choose the **Create Point Group** tool from the **Create Ground Data** panel; the **Point Group Properties - Point Group - (1)** dialog box will be displayed, as shown in Figure 2-6.

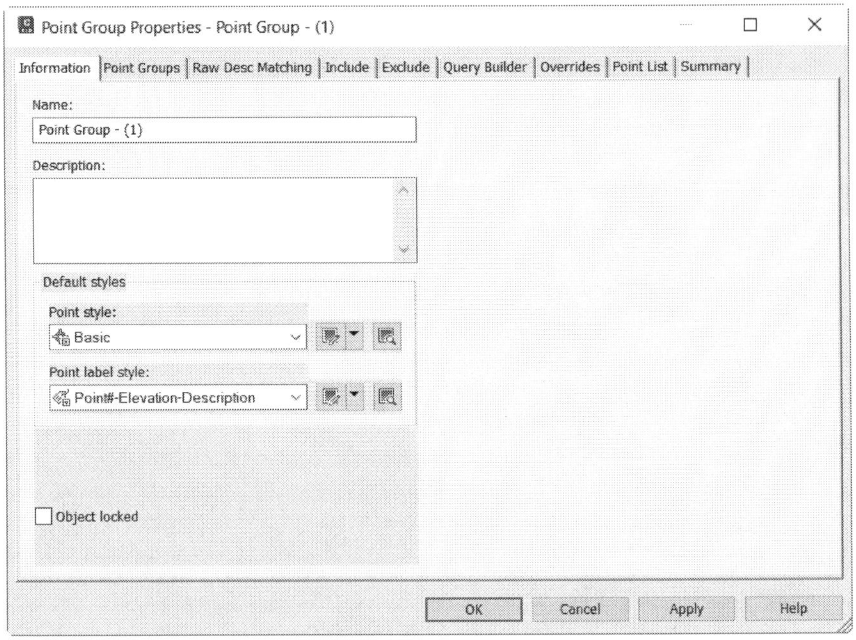

*Figure 2-6 The **Point Group Properties - Point Group - (1)** dialog box*

Editing Point Groups

To edit point group properties, right-click on the point group in the **Point Group** node in the **Prospector** tab and choose the **Properties** option from the shortcut menu; the **Point Group Properties - <point group name>** dialog box will be displayed. In the **Information** tab of the dialog box, you can modify the point group name, point style, point label style of points included in the point group. Choose the **Include** or **Exclude** tab from the dialog box to include and exclude points. Choose the **Apply** button to apply the changes and choose **OK** to close the dialog box.

TUTORIAL

Before starting the tutorial, you need to download and save the tutorial files on your computer. To do so, follow the steps given below:

1. Log on to *www.cadsofttech.com* and browse to *Books > Civil/GIS Books > AutoCAD Civil 3D > AutoCAD Civil 3D 2024 for Novices*. Next, select *c02_c3d_2024_tut.zip* file from the **Tutorial Files** drop-down list. Choose the **Download** button to download the data file.

2. Now, save and extract the downloaded folder to the following location:

   C:\c3d_2024

Note

*While opening the tutorial file, the **PANORAMA** window may appear with an error message. Close this window to proceed further.*

Tutorial 1 Creating Points I

In this tutorial, you will create points using various tools that are available in the **Create Points** toolbar, as shown in Figure 2-7. **(Expected time: 20 min)**

The following steps are required to complete this tutorial:

a. Open the file.
b. Create points using different options in the **Points** drop-down.
c. Save the file.

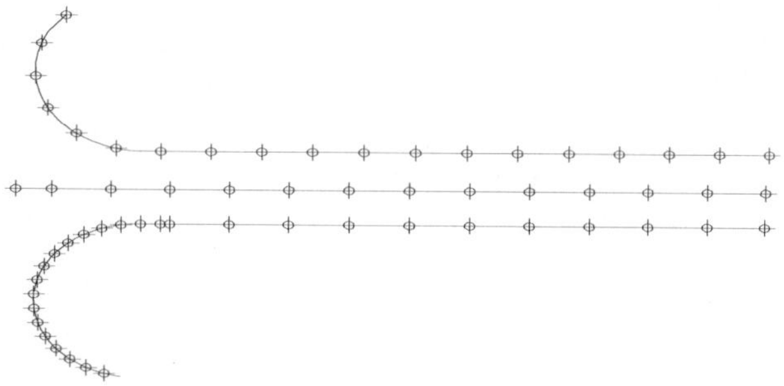

Figure 2-7 *Points created by using various tools in the* ***Create Points*** *toolbar*

Opening the File

1. Choose **Open > Drawing** from the **Application Menu**; the **Select File** dialog box is displayed.

2. In this dialog box, browse to the location *C:\c3d_2024\c02_c3d_2024_tut*.

3. Select the file *c02_c3d_2024_tut01.dwg* and then choose the **Open** button to open the file.

The opened drawing file consists of an alignment and two lines on both sides.

Creating Points Using the Options in the Alignment Category

1. Choose the **Create Points - Alignments** option from **Home > Create Ground Data > Points** drop-down; a flyout is displayed.

2. Choose the **Station/Offset** tool from the flyout to create points at some offset distance from the stations on an alignment; the **Create Points** toolbar is displayed and you are prompted to select the alignment.

3. Select the **Centreline** alignment in the drawing; you are prompted to specify the station along the baseline.

4. As you move the cursor over the alignment, a tooltip showing the first stations is displayed, as shown in Figure 2-8. Zoom in to view the alignment and click on **0+00** at the start of the alignment. Now you are prompted to specify the offset distance.

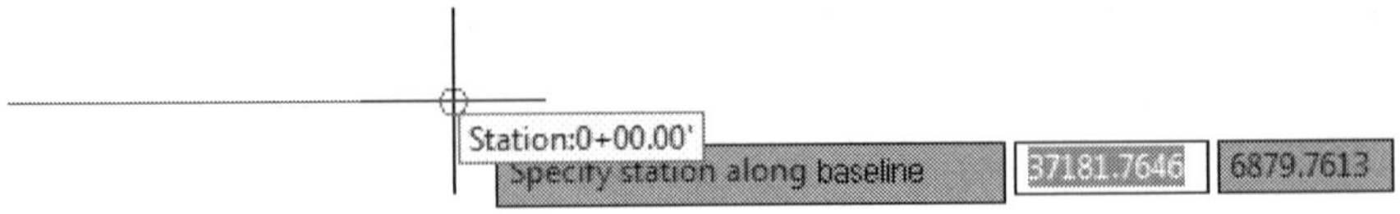

Figure 2-8 A tooltip showing the first station

Tip
*You can also create a point at the start of an alignment by specifying **0.00** in the command line and then press **ENTER**.*

5. Press ENTER to accept the default offset value; you are prompted to enter the point description.

6. Next, enter **CL** as the point description in the command line and press ENTER; you are prompted to specify elevation.

7. Specify the point elevation value as **100** in the command line, and press ENTER; a blue colored point is displayed at the first station of the alignment. Also, you are prompted to specify another station in the command line.

8. Pan to the next station **30+00** on the alignment and click on it.

9. Press ENTER to accept the default offset value; you are prompted to specify the description.

10. Repeat the procedure followed in steps 6 and 7 and add a new point at station **30+00**. On creating the point, you are again prompted to specify the next station.

Continue creating points till you reach the other end of the alignment, as shown in Figure 2-9.

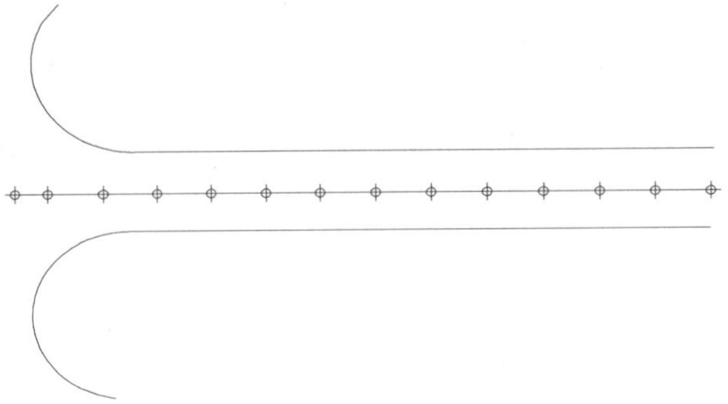

Figure 2-9 Points created at each selected station

11. Press ESC to terminate the command.

In this way, you have created points from the alignment.

> **Tip**
> *In case, the points do not appear in the drawing you can set the point type by entering the* ***DDPTYPE*** *command in the commandbar. As a result, the* ***Point Style*** *dialog box is displayed. In this dialog box, select the required type of point and then choose the* ***OK*** *button; the points are displayed in the drawing.*

Creating Points Using the Options in the Miscellaneous Category

1. Choose the **Create Points - Miscellaneous** option from **Home > Create Ground Data > Points** drop-down; a flyout is displayed.

2. Choose the **Divide Object** tool from the flyout; the **Create Points** toolbar is displayed and you are prompted to select an arc, line, polyline, lot line, or feature line.

3. Select the line above the alignment; you are prompted to enter the number of segments.

4. Enter **12** in the command line and press ENTER; you are prompted to specify the offset.

5. Press ENTER to accept the default offset value; you are prompted to specify the point description.

6. Enter **OSP** and press ENTER; you are prompted to specify the point elevation.

7. Enter **100** and press ENTER; a blue colored point is displayed at the endpoint of the selected line; you are prompted to specify the description and elevation.

8. Repeat the procedure given in steps 6 and 7 and keep on specifying the point description and elevation till you are prompted to select an arc, line, polyline, or feature line.

9. Next, select the arc object above the alignment; you are prompted to specify the number of segments. Enter **6** in the command line and press ENTER; you are prompted to specify the offset.

10. Enter **5** as the offset distance in the command line and press ENTER; you are prompted to specify the point description.

11. Enter **OSP** in the command line and press ENTER; you are prompted to specify the point elevation.

12. Enter **100** in the command line and press ENTER; you are prompted to specify the point description.

13. Repeat the procedure given in steps 11 and 12 to create points at an offset along the selected arc.

14. Press ESC to terminate the command.

Figure 2-10 shows points created from the line and arc objects by dividing these objects into specified segments.

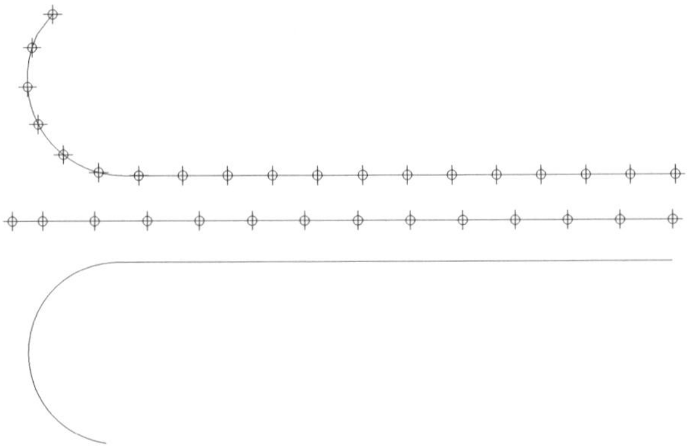

Figure 2-10 *Points created from the line and arc objects*

Creating Points Using the Options in the Intersection Category

1. Choose the **Create Points-Intersections** option from **Home > Create Ground Data > Points** drop-down; a flyout is displayed

2. Choose the **Direction/Perpendicular** tool from the flyout; you are prompted to specify the start point.

3. Click at the start point of the alignment to specify as the first point, refer to Figure 2-11; you are prompted to specify the direction.

4. Drag the cursor vertically downward and click on the line just below the point on alignment; an arrow showing the direction is displayed and you are prompted to specify the offset.

5. Press ENTER to accept the default offset value; you are prompted to specify the location for the perpendicular point.

6. Click on the line object just below the first point on the alignment; a green cross-mark indicating the position of the point is displayed on the screen. Also, you are prompted to specify the point description.

Note

*Ensure that the **OSNAP** or **ORTHO** command is activated to specify the perpendicular point. You can also use the transparent commands to specify the perpendicular points.*

7. Enter **PP** (perpendicular point) and press ENTER; you are prompted to specify the elevation.

8. Enter **100** in the command line and press ENTER; a point is displayed on the line and you are prompted again to specify the start point.

9. Now, click at the second point on the alignment; you are prompted to specify the direction.

10. Again, drag the cursor downward and click on the line just below the selected point; an arrow is displayed and you are prompted to specify the offset.

11. Press ENTER to accept the default offset distance; you are prompted to specify the location of the perpendicular point.

12. Click on the line just below the second point; you are prompted to specify the point description.

13. Specify the description and the elevation as specified in steps 7 and 8; a point is displayed on the line, refer to Figure 2-11.

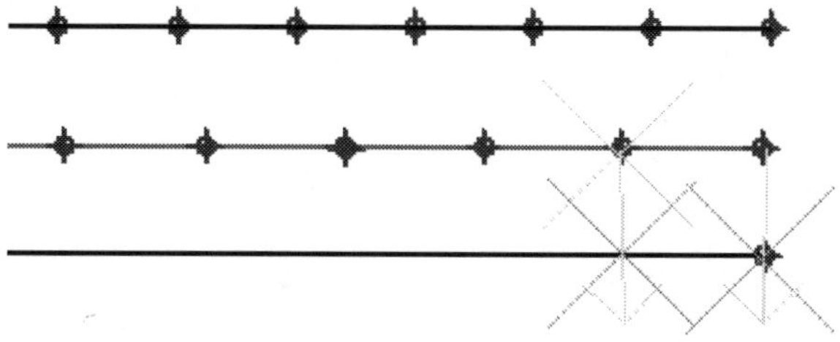

Figure 2-11 *Cross-marks indicating the location of the point*

14. Using the procedure followed in steps 3 to 8, create points in the perpendicular direction of the alignment points.

Figure 2-12 shows the points created at the intersection of the direction specified from alignment points and the perpendicular points specified on the line.

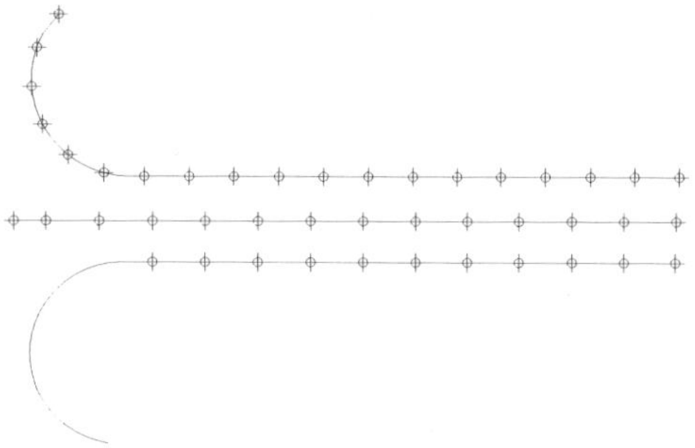

Figure 2-12 *Points created by intersection*

15. Choose the **Create Points - Miscellaneous** option from **Home > Create Ground Data > Points** drop-down; a flyout is displayed.

16. Choose the **Measure Object** tool from the flyout; you are prompted to select an arc, line, polyline, lot line, or feature line. Select the arc below the alignment; you are prompted to specify the starting station.

17. Press ENTER to accept the default value **<0.000>**. On doing so, the ending station is displayed in the command line.

18. Next, press ENTER; you are prompted to specify the offset distance.

19. Again, press ENTER; the default value is accepted as offset and you are prompted to specify the interval.

20. Enter **1000** in the command line and press ENTER; you are prompted to specify the point description.

21. Enter **OSP** in the command line and press ENTER; you are prompted to specify the elevation.

22. Enter **100** in the command line and press ENTER; you are again prompted to specify the description and elevation.

23. Repeat steps 21 and 22 till you are prompted to select an arc again. Thus, Civil 3D creates the point first by measuring the arc object by identifying the start and end stations of the arc and then divides the arc at specified intervals. Points are created at the end of each interval including the start and end stations of the arc, refer to Figure 2-13.

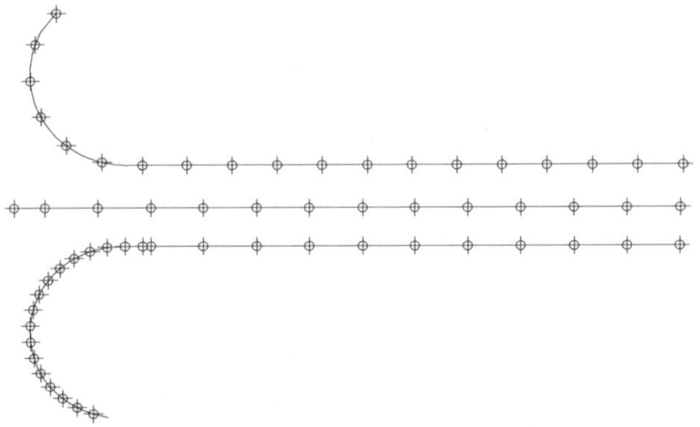

Figure 2-13 *Points created along the arc object below the alignment*

Saving the File

1. Choose **Save As** from the **Application Menu**; the **Save Drawing As** dialog box is displayed.

2. In this dialog box, browse to the following location:

   C:\c3d_2024\c02_c3d_2024_tut

3. In the **File name** edit box, enter **c02_tut01**.

4. Choose the **Save** button; the file is saved with the name *c02_tut01.dwg* at the specified location.

EXERCISES

Before starting the exercises, you need to download and save the exercise files on your computer. To do so, follow the steps given below:

1. Log on to *www.cadsofttech.com* and browse to *Books > Civil/GIS Books >AutoCAD Civil 3D > AutoCAD Civil 3D 2024 for Novices*. Next, select *c02_c3d_2024_exer.zip* file from the **Exercise Files** drop-down list. Choose the **Download** button to download the data file.

2. Now, save and extract the downloaded folder to the following location:

   C:\c3d_2024

Note

*While opening the exercise files, the **PANORAMA** window may appear with an error message. Close this window to proceed further.*

Exercise 1

Open the *c02_c3d_2024_ex01.dwg* file and create a point style and label style for points. Next, assign these styles to points using the following parameters: **(Expected time: 30 min)**

Point Style name: **New Style**
Marker Style: **default**
Marker Color: **blue**
Point Label Style Name: **New Label Style**
Color of all the label components: **blue**
Save the file as *c02_ex01a.dwg*.

Exercise 2

Using the *c02_ex01a.dwg* file, create a point group and edit the points in the point group using the **Point Editor**. Use the following parameters: **(Expected time: 30 min)**

Point Group Name: **Tower points**
Raw Description to include Points: **Tower**
Edit the elevation of the first five points, as shown in Figure 2-14
Save the file as *c02_ex02a.dwg*.

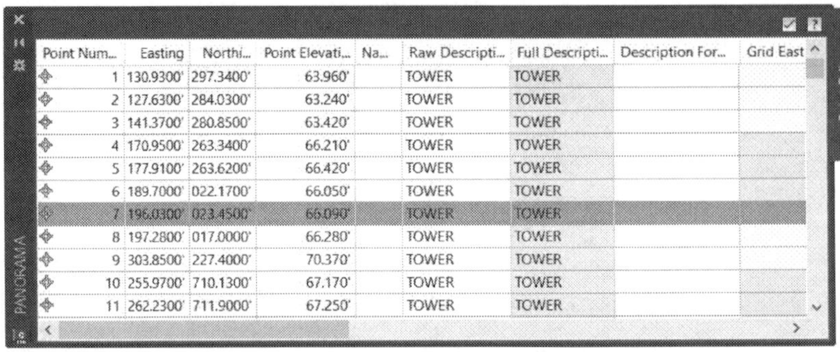

Point Num...	Easting	Northi...	Point Elevati...	Na...	Raw Descripti...	Full Descripti...	Description For...	Grid East
1	130.9300'	297.3400'	63.960'		TOWER	TOWER		
2	127.6300'	284.0300'	63.240'		TOWER	TOWER		
3	141.3700'	280.8500'	63.420'		TOWER	TOWER		
4	170.9500'	263.3400'	66.210'		TOWER	TOWER		
5	177.9100'	263.6200'	66.420'		TOWER	TOWER		
6	189.7000'	022.1700'	66.050'		TOWER	TOWER		
7	196.0300'	023.4500'	66.090'		TOWER	TOWER		
8	197.2800'	017.0000'	66.280'		TOWER	TOWER		
9	303.8500'	227.4000'	70.370'		TOWER	TOWER		
10	255.9700'	710.1300'	67.170'		TOWER	TOWER		
11	262.2300'	711.9000'	67.250'		TOWER	TOWER		

***Figure 2-14** The **Point Editor** window showing the edited elevations*

Chapter 3

Working with Surfaces

Learning Objectives

After completing this chapter, you will be able to:

- *Create various types of surfaces*
- *Add data from different sources*
- *Create and edit surface styles*

SURFACES

In Civil 3D, a surface is considered as a building block and is created from triangular networks or grid data. Each triangle on the surface is created by connecting three nearest points of elevation. Also, each triangle has a defined slope.

A surface is a basic requirement of any Civil 3D project. All volume calculations, profiles, corridors, sections, and grading objects that are generated in a project are based on the surface created. You can create a surface by importing data from Triangular Irregular Network (TIN) files, Digital Elevated Model (DEM) files, points, point files, existing AutoCAD objects, contours, breaklines, point groups, and also from Google Earth.

Types of Surfaces

There are four types of surfaces in Civil 3D: TIN Surface, TIN Volume Surface, Grid Surface, and Grid Volume Surface. The TIN Volume Surface and Grid Volume Surface are used for calculating volumes.

CREATING AND EDITING SURFACES

A surface can be created by using points, point groups, AutoCAD objects, and boundaries. A surface can also be created directly by importing the DEM and TIN files that are generated from the survey. The methods of creating a surface and its different aspects are discussed next.

Creating a Surface

Ribbon:	Home > Create Ground Data > Surfaces drop-down > Create Surface
Command:	CREATESURFACE

You can create a surface by using the **Create Surface** tool. Invoke the **Create Surface** tool from the **Create Ground Data** panel in the **Surfaces** drop-down; the **Create Surface** dialog box will be displayed, as shown in Figure 3-1.

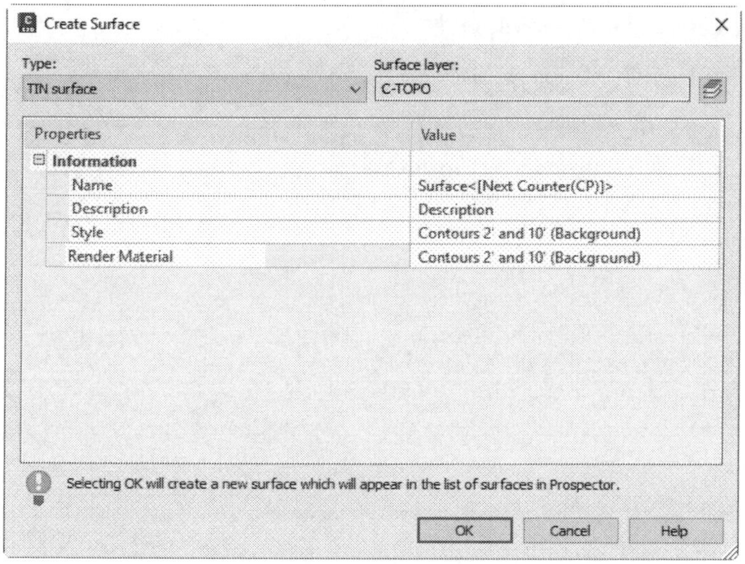

Figure 3-1 *The* ***Create Surface*** *dialog box*

Alternatively, you can invoke the **Create Surface** dialog box by using the **TOOLSPACE** palette. To do so, select the **Surfaces** node from the **Prospector** tab of the **TOOLSPACE** palette and right-click; a shortcut menu will be displayed. Choose the **Create Surface** tool from the shortcut menu to display the **Create Surface** dialog box. In this dialog box, you can specify the surface type, name, and description of the surface.

Adding Data from Different Sources to a Surface

You can add data to a surface from different sources such as points, point files, DEM and contours. The data obtained from these objects will be used to create a surface. The **Definition** subnode in the **Prospector** tab of the **TOOLSPACE** palette lists the different sources that can be used to add data to a surface. Expand the **Definition** subnode to view the sources, as shown in Figure 3-2.

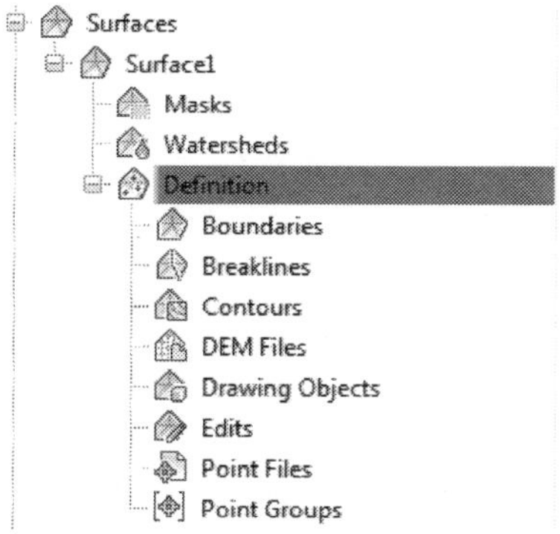

Figure 3-2 *Expanded* ***Definition*** *subnode*

Adding Surface Boundaries

Ribbon: Surface > Modify > Add Data drop-down > Boundaries
Command: ADDSURFACEBOUNDARIES

A boundary can be defined as a closed polygon entity used to represent the edge or limit of a surface. You can control the visibility of the surface triangulation within or outside the boundary. Any closed polygon or polyline object can be added to the surface as a surface boundary. To add a boundary to a surface, choose the **Boundaries** tool from the **Modify** panel; the **Add Boundaries** dialog box will be displayed, as shown in Figure 3-3.

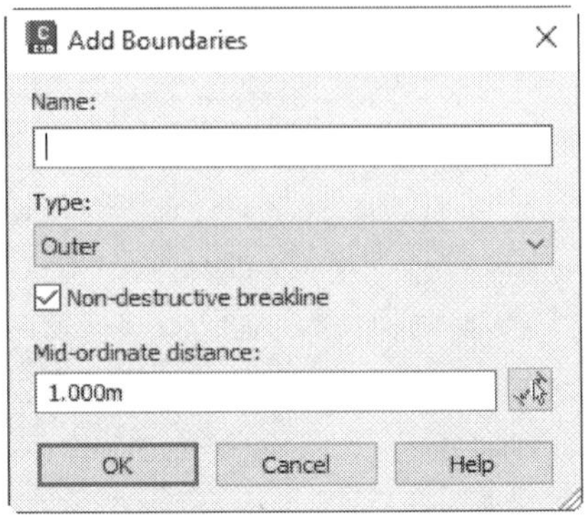

Figure 3-3 *The* ***Add Boundaries*** *dialog box*

Alternatively, you can add a boundary to a surface by using the options in the **TOOLSPACE** palette. To do so, expand the **Definition** subnode of the surface from the **Prospectors** tab and right-click on **Boundaries**; a shortcut menu will be displayed. Choose the **Add** option from the shortcut menu; the **Add Boundaries** dialog box will be displayed. In the dialog box, you can specify a name for the boundary in the **Name** edit box. Next, select the type of boundary from the **Type** drop-down list.

Deleting a Boundary

To delete a boundary, select it from the **TOOLSPACE** Item View and right-click to display a shortcut menu. Choose the **Delete** option from the shortcut menu; a warning message will be displayed, as shown in Figure 3-4. Choose the **OK** button; the boundary will be deleted from the drawing as well as from the **TOOLSPACE** Item View. Remember that the **Delete** option will not be displayed in the shortcut menu if you are trying to delete the current boundary. A small triangular symbol on the left of the boundary name in the **TOOLSPACE** item view indicates that a boundary is currently in use.

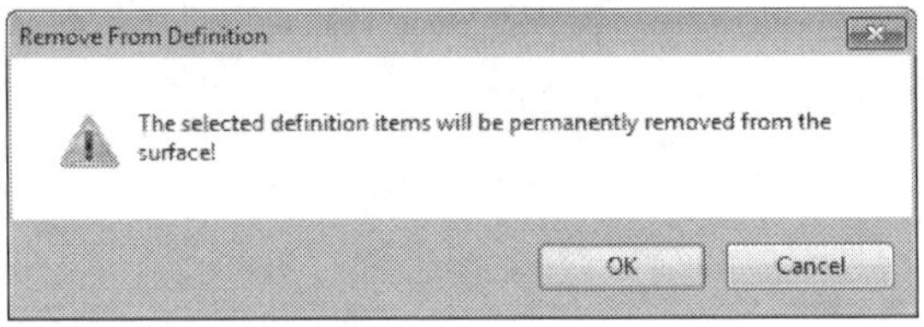

Figure 3-4 *The* ***Remove From Definition*** *warning message box*

Adding Breaklines to the Surface

Ribbon: Surface > Modify > Add Data drop-down > Breaklines
Command: ADDSURFACEBREAKLINES

Breaklines are polylines, feature lines, or 3D lines used to restrict the triangulation of the surface along the breakline. Breaklines basically represent the features such as streams, retaining walls, ditches, and so on. They can be added only to TIN surfaces. To add breaklines, choose the **Breaklines** tool from the **Modify** panel; the **Add Breaklines** dialog box will be displayed, as shown in Figure 3-5.

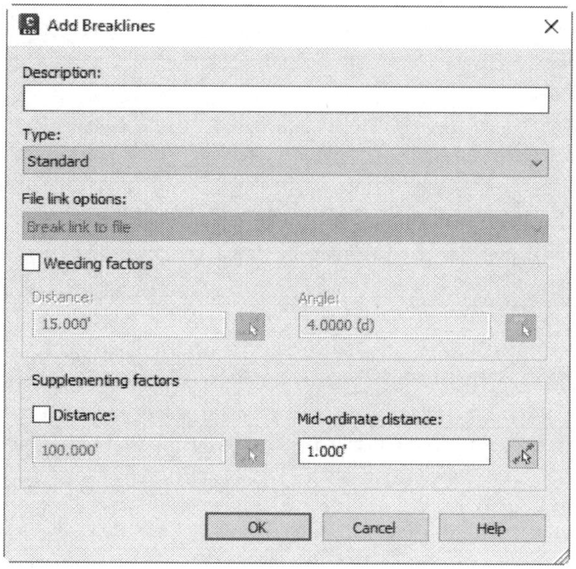

Figure 3-5 *The **Add Breaklines** dialog box*

Editing a Surface

You can perform various surface edits manually. To view various edit operations, choose the **Edit Surface** drop-down from the **Modify** panel; all the edit tools will be displayed. Choose any of the tools and perform the desired edit operation on the surface. Alternatively, you can edit surface by selecting the **Edits** option from the **Definition** subnode of the surface in the **Prospector** tab and right-click; a shortcut menu will display all surface edit options. The entire history of the edit operations performed on the surface will be added and can be viewed in the **TOOLSPACE** Item View.

Note

*The options in the shortcut menu displayed on choosing the Edits option of the **Prospector** tab are displayed based on the type of the surface selected.*

Creating Surface Masks

Surface masks are used to block the display of a surface or specific area on the surface and to assign it a different render material. The masking region can be defined by using closed polygons, polylines, 3D polylines, rectangles, feature lines, surfaces, and so on. To create a mask, expand the **Surfaces** node in the **Prospecto**r tab of the **TOOLSPACE** palette and then expand the required surface node. Next, right-click on **Masks** to display a shortcut menu and choose the **Create Mask** option from it; you will be prompted to select objects. Select the required polyline or other objects to define the boundary of the masking region and press ENTER; the

Create Mask dialog box will be displayed. Specify a name, masking type, render material, and other values in the dialog box and choose the **OK** button; the masked region will be created.

Creating Volume Surfaces

To create a volume surface, you need minimum two surfaces; base surface and comparison surfaces. Volume surfaces are classified into two types: TIN volume surfaces and Grid volume surfaces.

To create a volume surface, expand the **Surfaces** node in the **Prospector** tab of the **TOOLSPACE** palette and right-click; a shortcut menu will be displayed. Choose the **Create Surface** option from the shortcut menu; the **Create Surface** dialog box will be displayed. In this dialog box, select the **TIN volume surface** option from the **Type** drop-down list; the properties of the TIN volume surface will be displayed in the dialog box. Click in the **Value** field of the **Base Surface** property; a browse button will be displayed. Choose the browse button; the **Select Base Surface** dialog box will be displayed. Select the surface that is to be assigned as the base surface from the list of surfaces in the dialog box and choose the **OK** button. Similarly, click in the **Value** field of the **Comparison Surface** property and then select a surface to be assigned as the comparison surface from the **Create Surface** dialog box. Specify values of the other properties in this dialog box and choose the **OK** button; the dialog box will be closed and the volume surface will be created and listed in the **Surfaces** node. Similarly, you can also create grid volume surface using the **Create Surface** dialog box.

Note
Civil 3d will show error message box if surfaces overlap each other.

SURFACE STYLES

The surface styles control the display of the surfaces. You can create different types of surface styles as per your project requirements. The surface styles are created and managed by using the options in the **Surface Style** subnode in the **Settings** tab of the **TOOLSPACE** palette. Civil 3D also has some in-built surface styles that are listed in the **Surface Styles** node of the **Settings** tab. The surface styles that are listed in the **Surface Styles** node depend upon the type of template selected.

Creating a Surface Style

You can create a new surface style based on the project requirement. To create a new surface style, expand the **Surface** node in the **Settings** tab of the **TOOLSPACE** palette; the **Surface Styles** subnode will be displayed. Now, right-click on the subnode; a shortcut menu will be displayed. Choose the **New** option from the shortcut menu; the **Surface Style-New Surface Style** dialog box will be displayed, as shown in Figure 3-6.

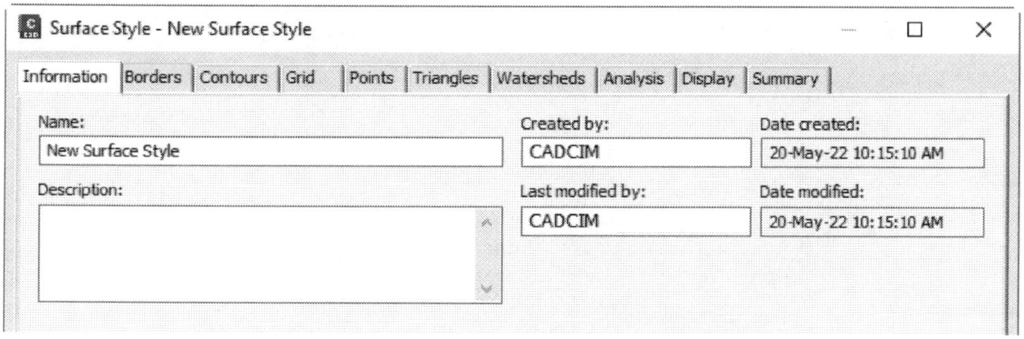

Figure 3-6 *Partial view of the **Surface Style - New Surface Style** dialog box*

Editing Surface Styles

Ribbon: Surface > Modify > Surface Properties drop-down > Edit Surface Style
Command: EDITSURFACESTYLE

To edit a surface style, choose the **Edit Surface Style** tool from the **Modify** panel; you will be prompted to select the surface. Select the required surface and press ENTER; the **Surface Style - <style name>** dialog box will be displayed. You can edit the properties of the surface style in this dialog box as per your requirement and choose the **OK** button to close it.

Tip
*Alternatively, choose the **Settings** tab in the **TOOLSPACE** palette and expand **Surface > Surface Styles** subnode. Select the required surface style and right-click on it; a shortcut menu will be displayed. Choose the **Edit** option from the shortcut menu; the **Surface Style - <style name>** dialog box will be displayed.*

SURFACE TOOLS

AutoCAD Civil 3D provides various tools which make the surface object more interactive. These tools can be accessed from the **Surface Tools** panel in the **Surface** contextual tab. To display the **Surface** tab, choose the **Surface** tool from the **Ground Data** panel of the **Modify** tab. The surface tools are discussed next.

Drape Image

Ribbon: Surface > Surface Tools > Drape Image
Command: DRAPEIMAGE

This tool is used to drape an image imported from Google Earth over an existing surface. Before using this tool, you need to import and insert an image from the Google Earth into Civil 3D and then overlay the image on the surface using the **Drape Image** tool.

To drape an image, choose the **Drape Image** tool from the **Surface Tools** panel; the **Drape Image** dialog box will be displayed, as shown in Figure 3-7.

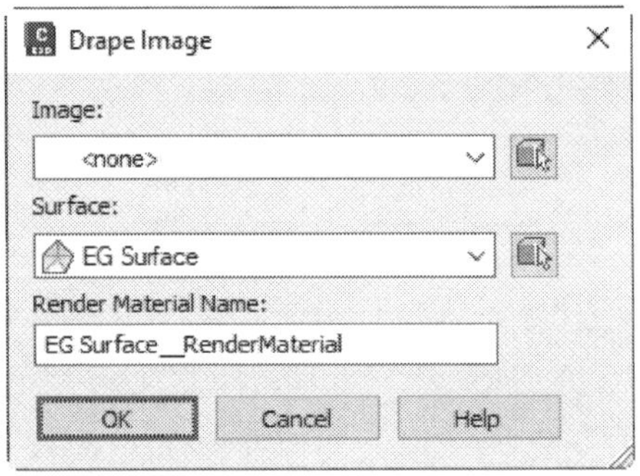

Figure 3-7 The **Drape Image** dialog box

Extract Solids from Surface

Ribbon:	Surface > Surface Tools > Extract from Surface drop down>Extract Solids from Surface
Command:	EXPORTSURFACETOSOLID

This tool is used to extract solids from the surface based on the surface triangles and surface borders. To extract a solid from a surface, choose the **Extract Solids from Surface** tool from **Surface Tools** panel; the **Extract Solid from Surface** dialog box will be displayed, as shown in Figure 3-8.

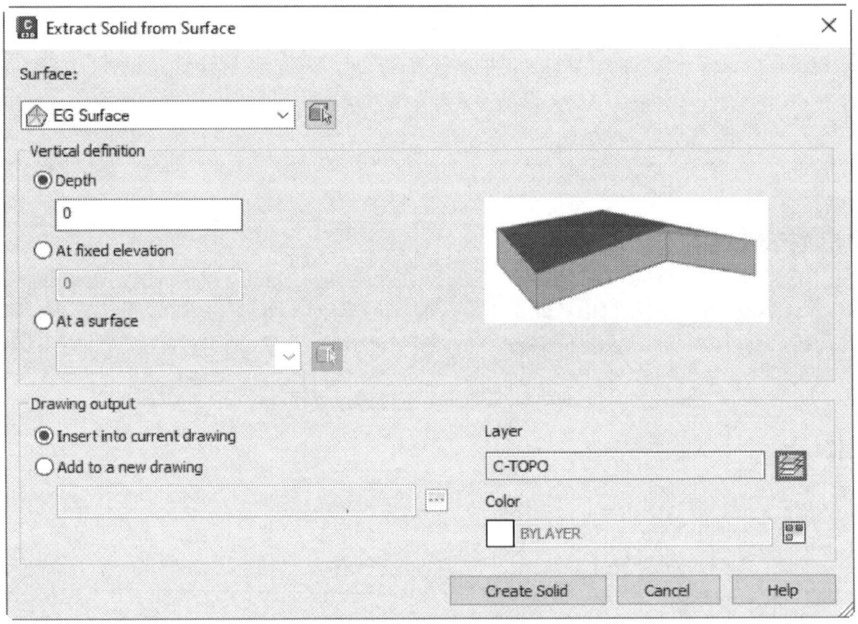

Figure 3-8 The **Extract Solid from Surface** dialog box

Export to DEM

Ribbon:	Surface > Surface Tools >Extract from Surface drop down>Export to DEM
Command:	SURFACEEXPORTTODEM

AutoCAD Civil 3D allows you to export the surface data to DEM files that can be used in different programs such as AutoCAD Map 3D, ArcGIS, and Raster Design. To export the surface data to a DEM file, choose the **Export to DEM** tool in the **Surface Tools** panel; the **Export Surface to DEM** dialog box will be displayed, as shown in Figure 3-9. This dialog box is used to specify the parameters to export a file. The **Selected Surface** category of this dialog box displays information of the surface to be exported and the **Export** category displays the properties of the DEM file to be created.

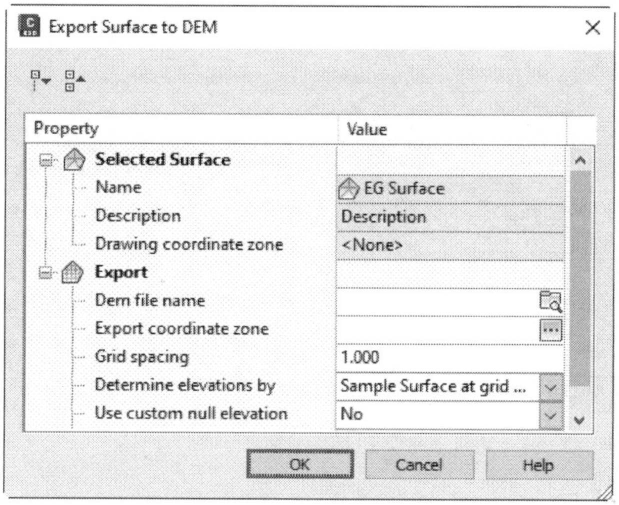

***Figure 3-9** The **Export Surface to DEM** dialog box*

Extract Objects

Ribbon:	Surface > Surface Tools >Extract from Surface drop down> Extract Objects
Command:	SURFACEEXTRACTOBJECTS

The **Extract Objects** tool is used to extract the surface components and modify them without exploding them. In such cases, the extracted components retain their original properties. These components can be modified and used for developing further designs. For example, catchment areas, water drop paths, watersheds, and so on can be extracted and the data obtained from these objects can be used to design ponds, sewers, and so on. These components can be used by the Hydrological department for further investigations.

To extract objects from a surface, choose the **Extract Objects** tool from the **Surface Tools** panel; the **Extract Objects from Surface - <surface name>** dialog box will be displayed, as shown in Figure 3-10.

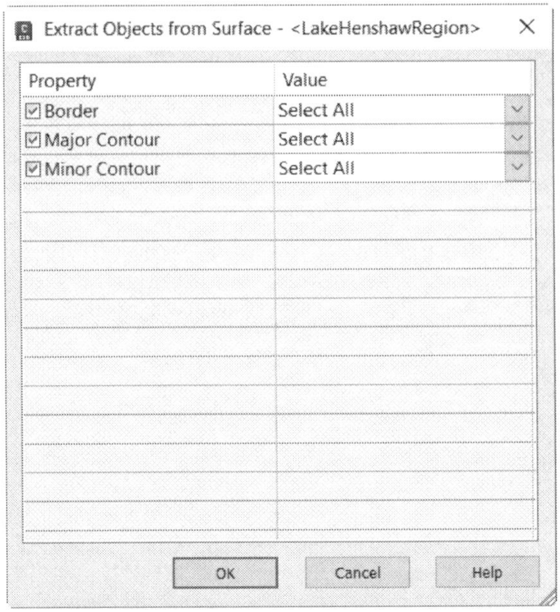

Figure 3-10 *The* ***Extract Objects from Surface - <EG Surface >*** *dialog box*

TUTORIAL

1. Download the *c03_c3d_tut.zip* file from *https://www.cadsofttech.com*. The path of the file is as follows: *Books > Civil/GIS Books > AutoCAD Civil 3D > AutoCAD Civil 3D 2024 for Novices.*

2. Now, save and extract the downloaded file at the following location:

   C:\c3d_2024

Note
*While opening the tutorial file, the **PANORAMA** window with an error message may appear. Close this window to proceed further.*

Tutorial 1 — Creating Surface

In this tutorial, you will create a surface, as shown in Figure 3-11, and add point data to it. Then, you will apply a surface style and perform edit operations on it. **(Expected time: 30 min)**

The following steps are required to complete this tutorial:

a. Open a template file.
b. Create a TIN surface.
c. Add Point File to the surface.
d. View the surface
e. Create and assign a new surface style to the surface.
f. Save the file.

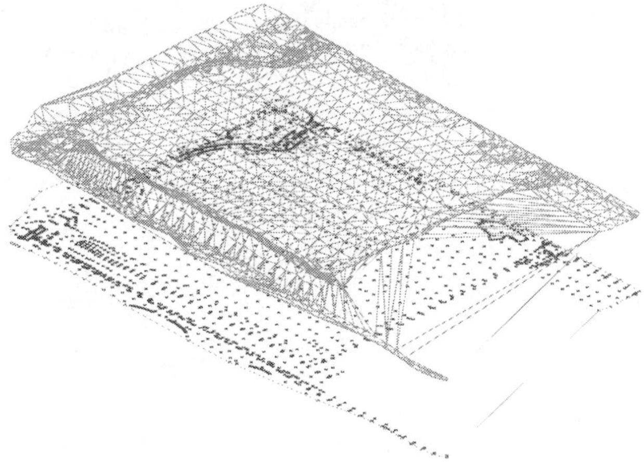

Figure 3-11 The surface displayed in 3D view

Opening the New Template File

1. Choose **New** from the **Application Menu**; the **Select template** dialog box is displayed.

2. In the **Select template** dialog box, select *_AutoCAD Civil3D (Imperial) NCS.dwt* template and choose the **Open** button.

Creating the Surface

1. Choose the **Create Surface** tool from **Home > Create Ground Data > Surfaces** drop-down; the **Create Surface** dialog box is displayed.

2. Click in the **Value** field of the **Name** property and choose the **Browse** button displayed on the right; the **Name Template** dialog box is displayed, as shown in Figure 3-12.

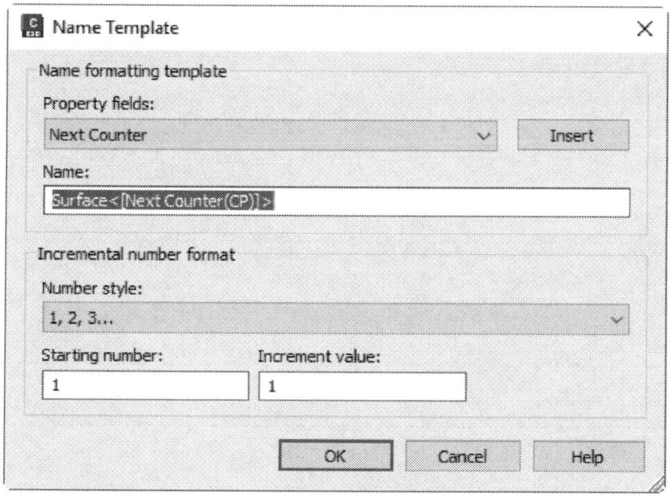

*Figure 3-12 The **Name Template** dialog box*

3. Enter the name **EG Surface** in the **Name** edit box of the dialog box and choose the **OK** button to close the dialog box.

4. Now, ensure that the **TIN surface** option is selected in the **Type** drop-down list of the **Create Surface** dialog box and choose the **OK** button; the dialog box is closed. Also, the surface is created and added to the **Surfaces** node of the **Prospector** tab of the **TOOLSPACE** palette, as shown in Figure 3-13.

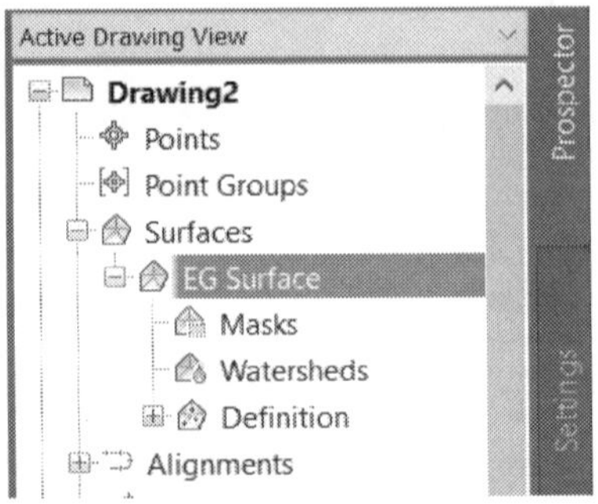

***Figure 3-13** The **EG Surface** added to the **Surfaces** node of the **Prospector** tab*

Note
You cannot see the surface at this stage because it is empty and you need to add data to make it visible.

Adding Point File to the Surface

1. Expand **Surfaces > EG Surface > Definition** in the **Prospector** tab of the **TOOLSPACE** palette.

2. Select **Point Files** from the **Definition** node and right-click; a shortcut menu is displayed.

3. Choose **Add** from the shortcut menu; the **Add Point File** dialog box is displayed.

4. From the **Specify point file format** list box, select the **PENZD (space delimited)** option. This file format is used for the point files that are saved with the extension *.txt*, as shown in Figure 3-14.

Note`
*If you select the wrong format for the point file from the **Specify point file format** list box, Civil 3D will display an error message indicating that the wrong file format has been selected.*

5. Next, choose the button on the right of the **Selected Files** list box; the **Select Source File** dialog box is displayed.

6. Browse to the folder *c03_c3d_2024_tut* and select the *c03 _c3d_2024_tut01- PENZD* .txt file. Choose the **Open** button; the file name and path is displayed in the **Selected Files** list box, refer to Figure 3-14.

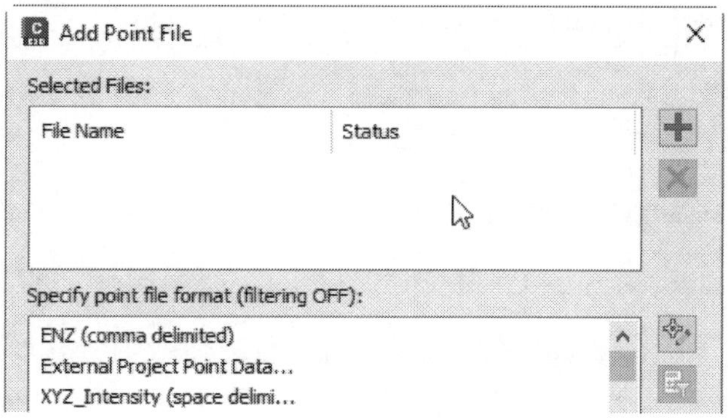

*Figure 3-14 Partial view of the **Add Point File** dialog box*

7. Choose the **OK** button from the **Add Point File** dialog box; the data is added to the surface.

Viewing the Surface

1. Enter **ZE** in the command line and then press ENTER; the surface is displayed with a surface border and the major and minor contours, as shown in Figure 3-15. This is due to the default surface style settings of the template file.

*Figure 3-15 The **EG Surface** displayed in the drawing area*

Creating a New Surface Style

1. Choose the **Settings** tab of the **TOOLSPACE** palette; the items in the tab are displayed.

2. In the **Settings** tab, expand **Surface > Surface Styles**; you will notice that an orange triangular symbol is displayed on the left of the three surface styles namely: **Contours 1' and 5' (Design)**, **Contours 2' and 10' (Background)**, and **Contours 2' and 10' (Design)** indicating that these surface styles are currently in use.

3. Select **Surface Styles** and right-click; a shortcut menu is displayed.

4. Choose **New** from the shortcut menu; the **Surface Styles - New Surface Style** dialog box is displayed. The **Information** tab is chosen by default in this dialog box.

5. Enter **EG Style** in the **Name** edit box of the **Information** tab.

6. Choose the **Apply** button; the name of the dialog box is changed to **Surface Style - EG Style**.

7. Next, choose the **Points** tab from the **Surface Style - EG Style** dialog box.

8. Click on the **Value** field of the **Point Units** property and enter **7.00'**.

9. Next, expand the **Point Display** category and click in the **Value** field of the **Data Point Symbol** property; a **Browse** button is displayed on the right of this field.

10. Choose the **Browse** button; a small window with different point symbols is displayed. Retain the default settings.

11. Next, choose the **Display** tab. Make sure that the **Plan** option is selected in the **View Direction** drop-down list of the **Display** tab. The settings of the surface style can only be viewed in the plan.

12. Click on the bulb icon in the **Visible** column and turn on the visibility of **Points** and **Triangles**. Also, ensure that the **Major Contour** and **Minor Contour** layers are turned on so that the objects in these layers are visible when viewed in the plan view.

13. Choose the **Apply** button to configure the settings for the plan view of the surface.

14. Next, select the **Model** option from the **View Direction** drop-down list.

15. In the **Display** tab, turn on the visibility of **Points** and **Border**. Also, ensure that the visibility of **Triangles** is turned on.

16. Choose the **Triangles** tab and click in the **Value** field of the **Triangle Display Mode** property and select the **Exaggerate Elevation** option from the drop-down list. Next, enter **5.000** in the **Value** field of the **Exaggerate Triangles by Scale Factor** property.

17. Choose the **Apply** button; the settings are configured for the 3D view/model of the surface.

18. Next, choose the **OK** button in the **Surface Styles - EG Style** dialog box; the **EG Style** is created and added to the **Surface Styles** node of the **Settings** tab.

Assigning the EG Style to the Surface

1. Choose the **Prospector** tab of the **TOOLSPACE** palette and expand the **Surfaces** node.

2. Select **EG Surface** and right-click; a shortcut menu is displayed.

3. Choose **Surface Properties** from the shortcut menu; the **Surface Properties - EG Surface** dialog box is displayed. The **Information** tab is chosen by default in this dialog box.

4. In the **Default styles** area of the tab, select **EG Style** from the **Surface style** drop-down list, as shown in Figure 3-16.

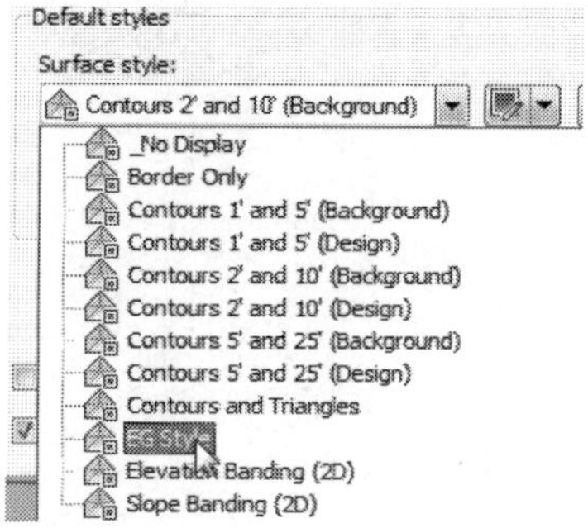

Figure 3-16 *Selecting the* ***EG Style*** *option from the* ***Surface style*** *drop-down list*

5. Choose the **Apply** button; a TIN surface is displayed with the surface points.

6. Choose the **OK** button to close the **Surface Properties - EG Surface** dialog box. Figures 3-17 and 3-18 show the TIN surface with the surface points and the point style selected, respectively. Note that the surface points are represented by the symbol that was selected for the point display in the surface style.

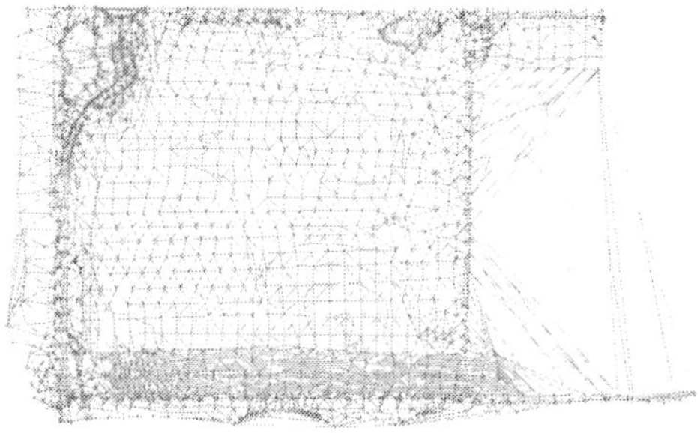

Figure 3-17 *The TIN surface with the surface points*

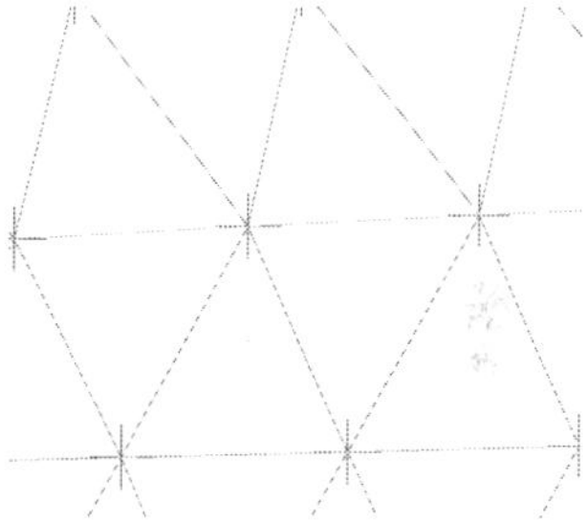

Figure 3-18 *Points represented by the selected point symbol*

7. Choose the **SE Isometric** button from the **Named Views** panel of the **View** tab, as shown in Figure 3-19 to view the surface in 3D view; the surface is displayed with triangles at an exaggerated elevation, and the surface border and points are displayed at the original elevation, as shown in Figure 3-20.

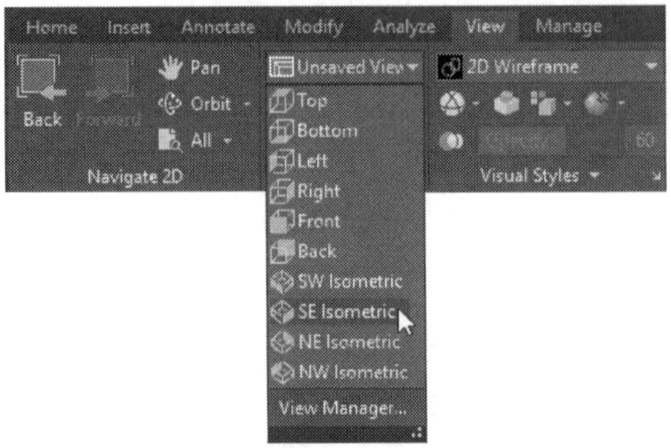

Figure 3-19 *The* ***SE Isometric*** *button chosen from the* ***View*** *tab*

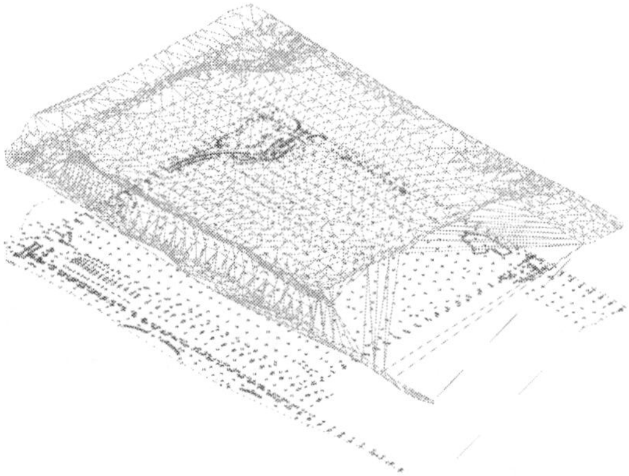

Figure 3-20 *The* ***SE Isometric*** *view of the* ***EG Surface***

Note
Figure 3-20 displays the model view of the surface using the model view settings previously defined for surface style.

Saving the Drawing

1. Choose the **Save As** from the **Application Menu**; the **Save Drawing As** dialog box is displayed.
2. In this dialog box, browse to the following location:

   C:\c3d_2024\c03_c3d_2024_tut

3. In the **File name** edit box, enter **c03_c3d_2024_tut01a**.
4. Next, choose the **Save** button; the file is saved with the name *c03_c3d_2024_tut01a* at the specified location.

EXERCISE

1. Log on to *www.cadsofttech.com* and browse to *Books > Civil/GIS Books > AutoCAD Civil 3D > AutoCAD Civil 3D 2024 for Novices*. Next, select *c03_c3d_2024_exer.zip* file from the **Exercise Files** drop-down list. Choose the **Download** button to download the data file.

2. Now, save and extract the downloaded folder to the following location:

   C:\c3d_2024

Note
*While opening the exercise files, the **PANORAMA** window may appear with an error message. Close this window to proceed further.*

Exercise 1 DEM Surface

Create a surface from *c03_c3d_2024_ex01.dem* file. Also, create a new surface style in the **Contours** tab of the **Surface Style - New Surface Style** dialog box using the following parameters:
(Expected time: 30 min)

1. Contour Display Mode: **Exaggerate Elevations**
2. Scale Factor: **3**
3. Minor Interval: **10**
4. Contour Smoothening: **True**
5. Smoothening type: **Spline curve**

Hint:
Settings in the **Display** tab

1. In **Plan** view direction, turn on the visibility of **Borders**, **Minor Contours**, and **Major Contours**.

2. Set the color of **Border** to **blue**, **Minor Contour** to **red**, and **Major Contour** to **green**.

3. In the **Model View** direction, turn off the visibility of **Border** and turn on the visibility of **Major Contour** and **Minor Contour** and set the color of **Minor Contour** to **magenta** and **Major Contour** to **cyan**.

Assign the newly created surface style to the surface and then choose the **SE Isometric** option from the list box in the **Views** panel of the **View** tab to view the surface. The SE Isometric view of the DEM Surface after assigning the new surface style is displayed, as shown in Figure 3-21.

Figure 3-21 *The SE Isometric view of the DEM Surface after assigning the new surface style*

Chapter 4

Surface Volumes and Analysis

Learning Objectives

After completing this chapter, you will be able to:

- *Understand the concept of surface volumes*
- *Compute surface volumes*
- *Understand the concept of surface analysis*
- *Perform surface analysis*
- *Create and add legend tables*

SURFACE VOLUMES

Earthwork estimation is an important part of any Civil Engineering project. Civil 3D provides you with a streamlined method to calculate total cut, fill, and net volumes of earthwork between two surfaces. The earthwork volumes between two surfaces can be calculated either by creating a volume surface (third surface) or by directly calculating the total cut and fill volumes between them.

Civil 3D makes earthwork calculations simple and fast. For earthwork calculations, you need two surfaces. Civil 3D analyzes both surfaces and calculates the total volume of cut and fill by comparing the vertical distance between points in these surfaces. The distance between the elevations of two surfaces can then be used to create a third surface, known as volume surface. The various methods to calculate the surface volumes are discussed next.

Volumes Dashboard

Ribbon: Analyze > Volumes and Materials > Volumes Dashboard
Command: VOLUMESDASHBOARD

Using the Volumes dashboard, you can break the surfaces in phases or parcels for volume analysis. The tools in the dashboard can be used to calculate and analyze the volume in multiple volume surfaces and bounded areas, which fall within the boundary of the volume surfaces. In the dashboard, you can save and use the information of the analysis across multiple drawing sessions. The result of volume analysis can be exported as a summary report or can be inserted into the drawing using the dashboard.

To calculate surface volumes, choose the **Volumes Dashboard** tool from the **Volumes and Materials** panel of the **Analyze** tab; the **PANORAMA** window with the **Volumes Dashboard** tab will be displayed, as shown in Figure 4-1.

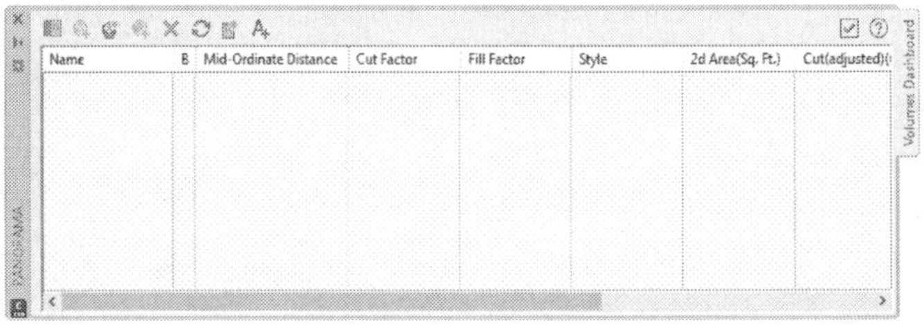

***Figure 4-1** The **PANORAMA** window with the **Volumes Dashboard** tab*

Composite Volumes

In the composite volume method, the cut, fill, and net volumes are calculated by using a pair of surfaces consisting of a top (Comparison) and a bottom (Base) surface. The Base surface is the existing ground surface and the Comparison surface will be the proposed surface. Civil 3D analyzes and compares the two surfaces to calculate the difference of elevation between each vertex point of the TIN or a cell in the Grid surfaces or at the intersection of the triangle edges of two surfaces. This elevation difference is then used to create a TIN volume or a TIN grid surface.

To use the composite volume method, type **reportsurfacevolume** in the command line; the **PANORAMA** window with the **Composite Volumes** tab will be displayed, as shown in Figure 4-2.

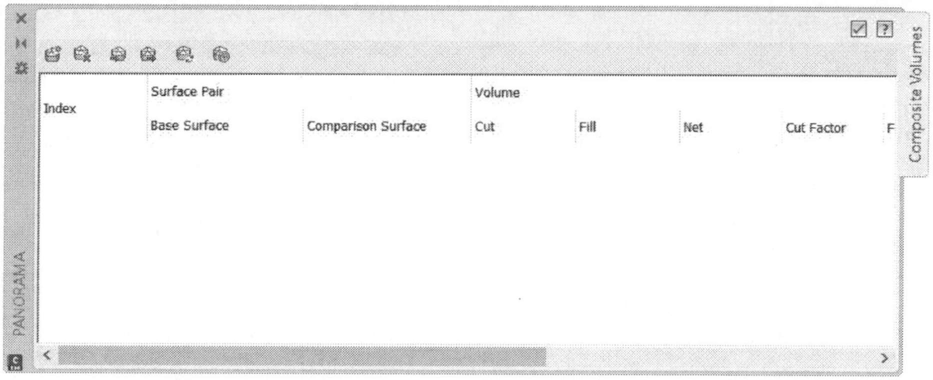

Figure 4-2 The ***PANORAMA*** *window with the* ***Composite Volumes*** *tab*

Bounded Volumes

This method helps you to calculate the volume of a surface or specified part of a surface. The area, whose volume has to be calculated, can be defined by a bounding polygon or a parcel. The bounded volume can be calculated using a volume surface or any other surface. In case of volume surfaces, the volume is calculated using the difference in elevation between two surfaces. For terrain surfaces, the volume is calculated from zero elevation to the elevation in the area bounded by a polygon or parcel.

CREATING VOLUME SURFACES

The volume surfaces help you review earthwork calculations of a site. The volume surface is a surface that is created by using the elevation difference between two surfaces. You can then analyze the surface and review the total amount of the cut or fill volume so that a balance between the total cut and fill volumes can be produced on the site. Civil 3D allows you to create two types of volume surfaces. The methods to create these volume surfaces are discussed next.

Creating the TIN Volume Surface

Ribbon:	Home > Create Ground Data > Surfaces drop-down > Create Surface
Command:	CREATESURFACE

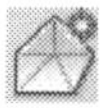

To create a TIN volume surface, you need a pair of TIN surfaces. In Civil 3D, you can create a TIN volume surface by choosing the **Create Surface** tool from the **Create Ground Data** panel, as discussed in the previous chapter. On doing so, the **Create Surface** dialog box will be displayed. Next, select the **TIN volume surface** option from the **Type** drop-down list in this dialog box. Specify the name, style, and description of the surface in the respective **Value** fields. Next, click in the **Value** field of the **Base Surface** property; a browse button will be displayed. Choose this button to display the **Select Base Surface** dialog box. In this dialog box, select the required option from the list box and then choose the **OK** button; the **Select Base**

Surface dialog box will be closed. Similarly, select the Comparison surface from the **Select Comparison Surface** dialog box. Specify the cut and fill factors for the material. Now, choose the **OK** button from the **Create Surface** dialog box; the dialog box will be closed and a TIN volume surface will be created and added to the **Surfaces** node. You can edit and analyze this surface the same way as you analyze to any other surface.

Note

*Due to increase in volume, the **Cut Factor** for the excavated material will be greater than 1 while the **Fill Factor** will be less than 1 as the volume of the filling material decreases due to compaction.*

Creating the Grid Volume Surface

Ribbon: Home > Create Ground Data > Surfaces drop-down > Create Surface
Command: CREATESURFACE

You can create a grid volume surface by using a pair of grid surfaces. To create a Grid volume surface, first you need to create a grid surface, as discussed in previous chapter. Next, choose the **Create Surface** tool from the **Create Ground Data** panel; the **Create Surface** dialog box will be displayed. Select **Grid volume surface** from the **Type** drop-down list; the properties of the grid volume surface will be displayed in the **Properties** column of the dialog box. Specify the information such as name, description, and style for the grid volume surface in the respective **Value** fields. Similarly, you can specify the grid parameters such as grid spacing in both X and Y directions and also in the grid's orientation. Next, select the Base surface as well as the Comparison surface as explained in the previous section. Enter the cut and fill factors and then choose the **OK** button. The **Create Surface** dialog box will be closed and the grid volume surface will be added in the **Surfaces** node.

SURFACE ANALYSIS

Ribbon: Surface > Modify > Surface Properties drop-down > Surface Properties
Command: EDITSURFACEPROPERTIES

After creating a surface, you need to analyze it for collecting information from it. This information can be used in the development of the project design. Surface analyses helps you understand and visualize the surface topography in a better way. For example, if you are generating the slope analyses of a surface, it would help you to find slopes of various points in the surface. Civil 3D supports seven ways to analyze a surface. These are Contours, Directions, Elevations, Slopes, Slope Arrows, Watersheds, and User-defined contours.

Contour Analysis

The **Contours** option in the **Analysis type** drop-down is used to analyze and display surface contours by using different colors and linetypes. To analyze contours, select the **Contours** option from the **Analysis type** drop-down list of the **Surface Properties - <surface name>** dialog box. The **Legend** area displays the default legend style for the selected analysis type. Civil 3D has an in-built legend style for each of its analysis types. However, you can create your own legend style and display the required information in the legend. This is discussed later in this chapter. In the **Analysis** tab, you can also preview the legend style. To preview the legend tables with analysis information, select the **Preview** check box.

Direction Analysis

The **Directions** option in the **Analysis type** drop-down list of the **Analysis** tab is used to classify surface triangles on the basis of the direction they face. Direction analysis generates aspect map that graphically displays the surface slopes with reference to the cardinal direction. For analyzing the surface, select the **Directions** option from the **Analysis type** drop-down list of the **Surface Properties - <surface name>** dialog box. Specify the options in the **Legend** area, as discussed earlier. In the **Ranges** area, specify the value in the **Number** spinner and choose the **Run analysis** button; the result of this analysis will be displayed in terms of maximum and minimum directions of each range with the assigned color scheme in the **Range Details** area of the **Analysis** tab.

Elevations Analysis

The Elevations analysis is the most important type of Surface analysis. It is used to group the surface triangles based on their elevation. It helps the site planners review and judge the total amount of earthwork required to create a leveled finished surface. To run this type of analysis, select the **Elevations** option from the **Analysis type** drop-down list in the **Analysis** tab of the **Surface Properties - <surface name>** dialog box. Specify the options in the **Legend** area as discussed earlier. In the **Ranges** area, select the required option from the **Create ranges by** drop-down list. You will need to set the value in the spinner, if you select the **Number of ranges** option from the drop-down list. Next, run the analysis; the details such as maximum elevation, minimum elevation, and color scheme for each range of elevation will be displayed in the **Range Details** area of the **Analysis** tab. Also, elevation bands will be displayed in different shades of color specified in the **Color Scheme** property of the **Analysis** tab in the **Surface Style** dialog box. You can also modify the color scheme in the **Surface Style** dialog box. Note that the default color scheme selected is **Blue**.

Slopes Analysis

The Slopes analysis is another important analysis carried out to analyze varying slopes in the surface. The slope of a surface is defined as a change in its elevation over a certain distance. It can be expressed as a ratio or a percentage (grade). For example, slope of 4:1 means an elevation raise of one unit for a distance of four units. In percentage, it can be expressed as 25% of grade. Similarly, a slope of 3: -1 means an elevation fall of one unit for a distance of three units. In percentage, it can be expressed as a grade of -33%. The **Slopes** option in the **Analysis type** drop-down is used to display the slope of the ground. In Slope analysis, the surface triangles are grouped according to their slope. Each range is assigned a color and the triangles in a particular slope based on range will be displayed using the same color scheme.

Slope Arrows Analysis

The **Slope Arrows** option in the **Analysis type** drop-down list is used to perform Surface analysis on the basis of the directions of slopes of the triangles. The result of this analysis is displayed using the direction arrows. These arrows show the slope direction and are placed at the centroid of each triangle in a TIN surface. The slope arrows pointing each other represent a valley whereas the arrows pointing away from each other signify a hill. In the Watershed analysis, the slope arrows represent the direction of the water flow across the surface.

User-defined Contours Analysis

The **User-defined contours** option in the **Analysis type** drop-down is used to carry out the analysis for the user-defined contour lines of the surface. These contour lines are defined according to their elevation ranges. For example, consider that the site on which you are working has contour lines upto 100 feet of elevation and your project requires analyzing contours upto 70 feet. In such a case, you can use the User-defined contours analysis. This analysis divides the surface and displays contour lines at specified contour intervals upto the elevation range defined by the user. As a result, in the example discussed above, after performing this analysis, the surface will display contour lines upto a range of 70 feet at specific contour intervals. Moreover, the contour lines are displayed in different colors, linetype, and lineweight.

Watersheds Analysis

Watershed is an area that is drained by watercourse. It is the area of land that catches rain, snow, and other forms of precipitation, and then drains it to another watercourse such as river, lake, and so on. The TIN lines of the surface are used to calculate the areas through which the water would flow along the surface. These areas help you determine the drain targets. Drain target is an area from where the water flow stops or passes over the surface. The region of the surface from where water drains to these targets is called a watershed.

In Civil 3D, the Watersheds analysis helps you analyze the flow of water along and off the surface by determining the regions or watersheds.

SURFACE ANALYSIS TABLES

In Civil 3D, data collected after performing a Surface analysis is divided into ranges that have different colors assigned to them. The surface legend table is used to organize and consolidate the data in a tabular form. You can create surface legend tables for all types of surface analyses, such as **Directions**, **Elevations**, **Slopes**, **Slope Arrows**, **Contours**, and **Watersheds**.

You can add a surface legend table to your drawing only after you have performed the Surface analysis, modified the surface style to display the analysis type, and set up a surface legend table style to display the surface data. The method of creating a legend table style and adding a legend table to the drawing is discussed next.

Creating Legend Table Styles

Civil 3D offers a default legend table style for each type of Surface analysis. However, you can create your own legend table style and view the required information in the table.

The **Table Styles** are created, edited, and managed in the **Settings** tab of the **TOOLSPACE** like any other Civil 3D object. To create a new legend table style, expand the **Surface > Table Style** node in the **Settings** tab and then expand the subnode for which you want to create a style. For example, expand the **Elevation** node to create the legend style for the Elevation analysis of the surface. Next, you will notice that a default **Elevations** legend style name appears in the **Elevation** node. Now, right-click on **Elevation** and then choose **New** from the shortcut menu; the **Table Style - New Elevation Table Style** dialog box will be displayed, as shown in Figure 4-3.

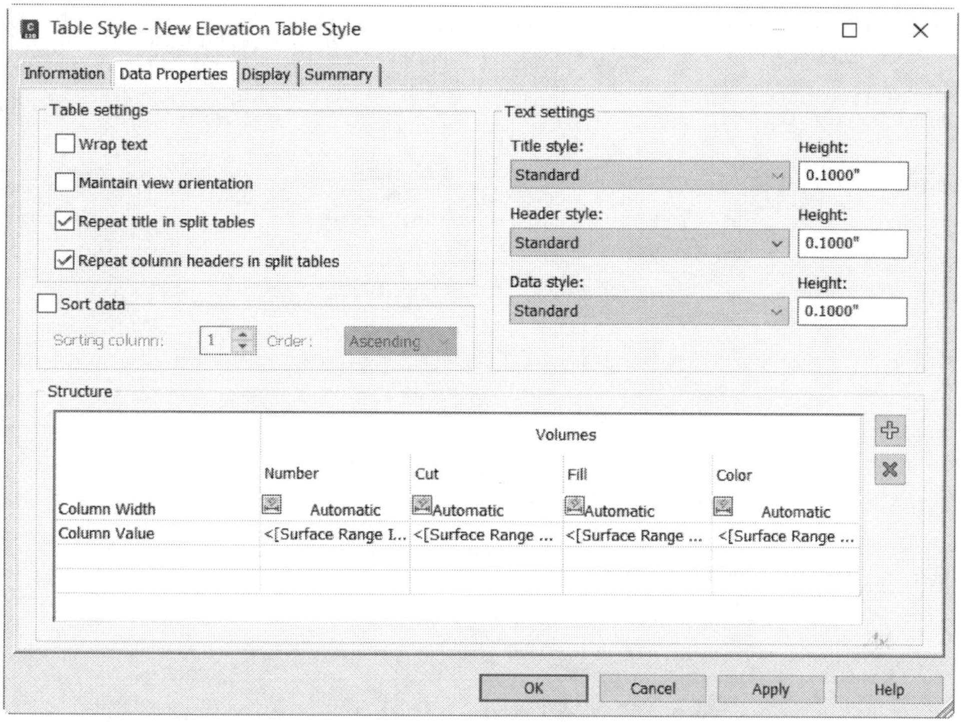

Figure 4-3 *The* ***Table Style - New Elevation Table Style*** *dialog box*

Adding Surface Legend Tables

Ribbon:	Annotate > Labels & Tables > Add Tables drop-down > Add Surface Legend Table
Command:	ADDSURFACELEGENDTABLE

To add the result table of Surface analysis in the current drawing, choose the **Add Surface Legend Table** tool from the **Labels & Tables** panel; you will be prompted to enter the table type. Specify the type of table by entering a value in the command line or by selecting it dynamically from the dynamic input prompt; you will be prompted to specify the behavior of the table. Select the **Dynamic** option so that the legend table is updated automatically when you modify the surface. If you select the **Static** option, the legend table will not be updated automatically when you modify the surface. On selecting the behavioral option for the legend table, you will be prompted to specify the upper left corner of the table. Click in the drawing to specify the upper left corner; the legend table will be added in the drawing.

Note

While adding the legend table if there are multiple surfaces, you will be prompted to select the surface or press ENTER; the ***Select a Surface*** *dialog box will be displayed with the name of the surfaces available. Select the required surface from the list and choose the* ***OK*** *button to close the dialog box.*

To assign a new table style to the added legend table, select the table from the drawing and right-click; a shortcut menu will be displayed. Choose **Table Properties** from the shortcut menu; the **Table Properties** dialog box will be displayed, as shown in Figure 4-4.

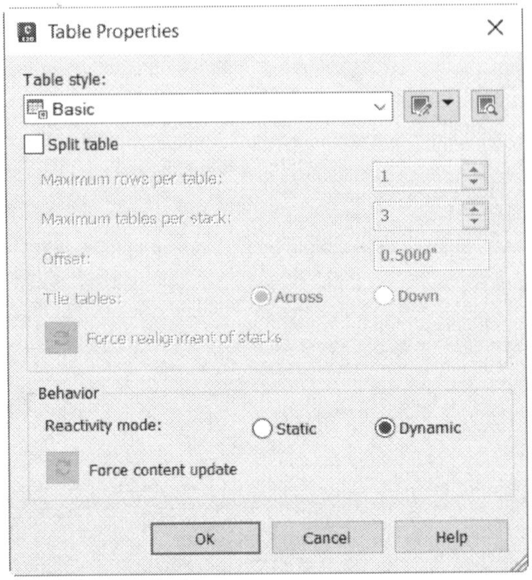

***Figure 4-4** The **Table Properties** dialog box*

SURFACE LABELS

Ribbon:	Annotate > Labels & Tables > Add Labels drop-down > Surface > Add Surface Labels
Command:	ADDSURFACELABELS

After you have created and analyzed the surface, you can display the surface information by using surface labels. Surface labels have two components: labels and tag. The label component contains text and the tag component is used to mark surfaces with numbers. Surface labels are created and managed in the **Settings** tab of the **TOOLSPACE** palette. You can add labels for surface contours, slopes, and spot elevations by using the **Add Surface Labels** tool. The various label types are discussed next.

Adding Contour Labels

To add contour labels to a surface, choose the **Add Surface Labels** tool from the **Labels & Tables** panel; the **Add Labels** dialog box will be displayed. Select **Surfaces** from the **Feature** drop-down list and then select **Contour - Multiple at Interval** from the **Label type** drop-down list. Next, select the major contour label, minor contour label, and user contour label styles from the respective drop-down lists. Now, choose the **Add** button from the dialog box; you will be prompted to pick the first point. Click in the drawing to specify the first point of the line along which you want to label contours. Click again to specify the second point of the line. Next in the command line, specify the contour interval at which you want to label contours. Press ENTER; Civil 3D will label contours along the line at specified contour. Similarly, you can label multiple contours or a single contour by selecting the **Contour - Multiple** or **Contour - Single** option from the **Label type** drop-down list. Alternatively, you can label contours by choosing the **Contour - Multiple**, **Contour - Single**, or **Contour - Multiple at Interval** tool from the **Add Labels** drop-down, refer to Figure 4-5.

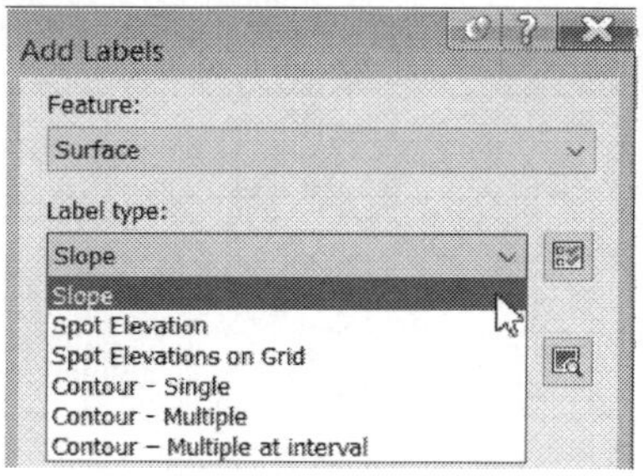

Figure 4-5 *Various options for labeling a surface*

Adding Slope Labels

You can add slope labels to a surface. To do so, choose the **Add Surface Labels** tool from the **Labels & Tables** panel; the **Add Labels** dialog box will be displayed. Select the **Surface** option from the **Feature** drop-down list in the dialog box. Next, select **Slope** from the **Label type** drop-down list.

Adding Spot Labels

Civil 3D allows you to add spot elevation labels and view the elevation of the surface at the spot. To add spot elevation labels, select the **Spot Elevation** option from the **Label type** drop-down list in the **Add Labels** dialog box. Select the required spot elevation label style option from the **Spot elevation label style** drop-down list. Also, select a marker style from the **Marker style** drop-down list to add a marker at the spot elevation label. Now, choose the **Add** button; you will be prompted to select the surface. Select the surface; you will be prompted to select a point in the drawing. Select the point in the drawing; the spot elevation along with the marker will be placed in the drawing for the selected point. Now, choose the **Close** button, the **Add Labels** dialog box will be closed. You can also use the **Spot Elevation on Grid** option from the **Label type** drop-down list in the **Add Labels** dialog box to add the spot elevation in a grid surface.

SURFACE LABEL STYLES

Surface styles control the display of surface, while label styles control the display of surface label styles. Civil 3D has various in-built label styles but you can also create your own label styles to label surface contours, slopes, watersheds, and spot elevations. These label styles control the visibility and display of surface labels. To create a surface label style, choose the **Settings** tab of the **TOOLSPACE** palette. Expand the **Label Styles** subnode from the **Surface** node in the **Settings** tab. Next, expand the required surface component node for which you want to create a label style. For example, in the **Contour** subnode, you will notice some in-built contour label styles. Now, if you want to create a new label style for contours, right-click on the **Contour** subnode to display a shortcut menu. Choose **New** from the shortcut menu; the **Label Style Composer - New Surface Contour Label Style** dialog box will be displayed, as shown in Figure 4-6.

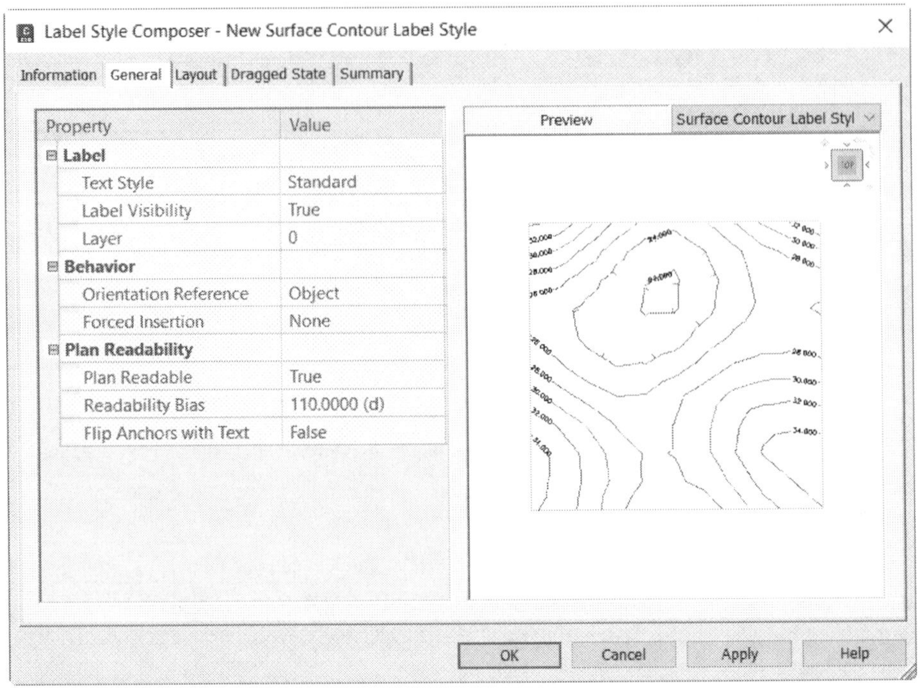

Figure 4-6 The ***Label Style Composer - New Surface Contour Label Style*** *dialog box*

TUTORIAL

1. Download the *c04_c3d_2024_tut.zip* file from *https://www.cadsofttech.com*. The path of the file is as follows: *Books > Civil/GIS Books > AutoCAD Civil 3D > AutoCAD Civil 3D 2024 for Novices.*

2. Save and extract the downloaded file at the following location:

   C:\c3d_2024

Note

While opening the tutorial file, the ***PANORAMA*** *window with an error message may appear. Close this window to proceed further.*

Tutorial 1 Surface Volume

In this tutorial, you will calculate the composite volume and create a TIN volume surface. Also, you will calculate the volume of an area bounded by a polygon. Figure 4-7 shows the surface whose volume has been determined. **(Expected time: 45 min)**

The following steps are required to complete this tutorial:

a. Open the file.
b. Compute composite volume.
c. Create a TIN volume surface.

d. Create the new legend style.
e. Add the legend table to the drawing.
f. Calculate the volume of the area bounded by the polygon.
g. Save the file.

Cut and Fill Levels				
Number	Cut Level	Fill Level	Color	Net Volume
1	−13.012	0.000	■	20567.31

Figure 4-7 *Surface with the determined net volume*

Opening the File

1. Choose **Open** from the **Application Menu**; the **Select File** dialog box is displayed.

2. In this dialog box, browse to the location *C:\c3d_2024\c04_c3d_2024_tut,* where you have saved the file.

3. Select the file *c04_c3d_2024_tut01.dwg* and choose the **Open** button to open it.

   The drawing file consists of two TIN surfaces: the **Existing Ground** surface with borders and contours, and the **Proposed Ground** surface with triangles. You will notice a magenta colored polygon in the **Proposed Ground** surface.

Computing Composite Volumes

1. Enter **REPORTSURFACEVOLUME** at the command prompt and press ENTER; the **Composite Volumes** tab in the **PANORAMA** window is displayed.

2. Choose the **Create new volume entry** button from the **Composite Volumes** tab of the **PANORAMA** window; a new volume entry is added to the list box. 

3. Double-click on the **<select surface>** cell in the **Base Surface** column; a drop-down list is displayed. Select the **Existing Ground** option from this drop-down list, as shown in Figure 4-8.

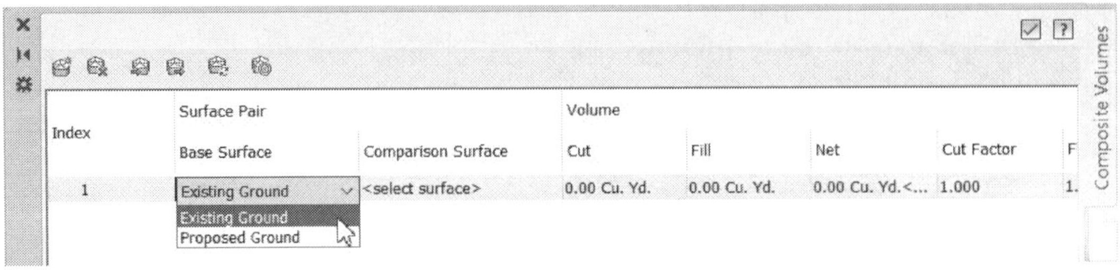

*Figure 4-8 Selecting the **Existing Ground** option from the drop-down list*

4. Similarly, select the **Proposed Ground** option from the drop-down list in the **Comparison Surface** column. On specifying the base and comparison surfaces, the net cut and fill volumes are displayed in the **Cut** and **Fill** columns, respectively, as shown in Figure 4-9.

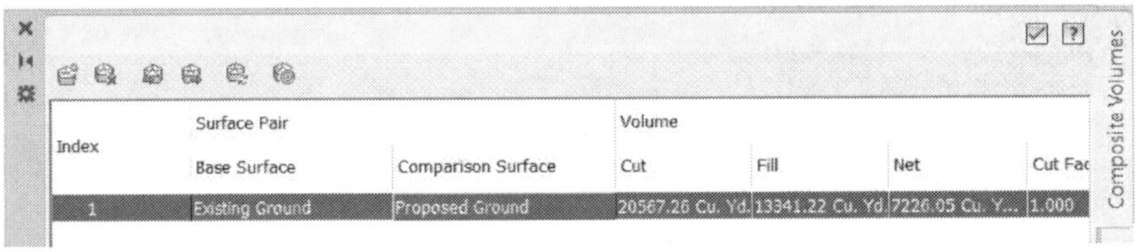

*Figure 4-9 The net cut and fill volumes displayed in the **Cut** and **Fill** columns*

The total volume of material to be removed from the existing ground is **20567.26 Cu. Yd.** and the total material required to be filled is **13341.22 Cu. Yd**.

5. Accept the default cut and fill factor values. Move the slider of the **PANORAMA** window to pan across the window to view the adjusted values for the cut, fill and net material required. The net volume of material is **7226.05 Cu Yd<Cut>**.

   This means that to create a balanced proposed ground, the **7226.05 Cu Yd** of material needs to be removed from the existing ground. Note that the red graph in the **Net Graph** column indicates that the material needs to be removed from the base surface.

6. Close the **PANORAMA** window.

Creating the TIN Volume Surface

1. Choose the **Create Surface** tool from **Home > Create Ground Data > Surfaces** drop-down; the **Create Surface** dialog box is displayed.

2. In this dialog box, select the **TIN volume surface** option from the **Type** drop-down list.

3. Set the values of the TIN volume surface properties as follows:

   Name: **Volume Surface**
   Base Surface: **Existing Ground**
   Comparison Surface: **Proposed Ground**

4. Choose the **OK** button; the **Create Surface** dialog box is closed. The surface is created and displayed in different colored bands.

5. Now, expand the **Surfaces** node in the **Prospector** tab of the **TOOLSPACE** palette.

6. Right-click on the **Volume Surface** node; a shortcut menu is displayed. From the shortcut menu, choose the **Surface Properties** option; the **Surface Properties - Volume Surface** dialog box is displayed.

7. Choose the **Statistics** tab from this dialog box if not selected by default.

8. Expand the **Volume** category in this tab; the cut, fill, and net volumes are displayed, as shown in Figure 4-10.

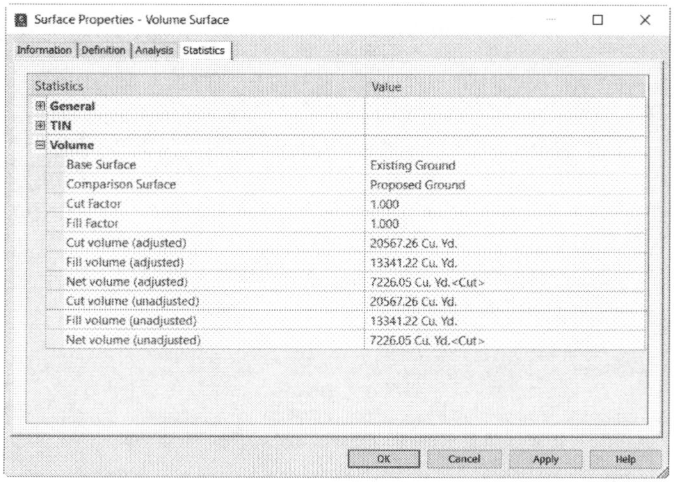

***Figure 4-10** The cut, fill and net volumes displayed in the **Volume** category*

Next, you will perform the Elevations analysis to visually differentiate various ranges of altitude in the surface.

9. Next, choose the **Analysis** tab in the **Surface Properties - Volume Surface** dialog box.

10. Ensure that the **Elevations** option is selected from the **Analysis type** drop-down list.

11. In the **Ranges** area, retain the default settings and choose the **Run analysis** button; the results are displayed in the **Range Details** area. Choose the **Apply** button; the volume surface shows elevation bands in different colors, as shown in Figure 4-11. Different color bands represent various elevation ranges in the surface.

***Figure 4-11** The volume surface with elevation band*

Now perform analysis using two bands of elevation ranges.

12. Set the value in the spinner located next to the **Create ranges by** drop-down list to **2** and choose the **Run analysis** button; the analysis is performed. The range details are displayed in the **Range Details** area, as shown in Figure 4-12. Choose the **Apply** button to apply the changes to the drawing.

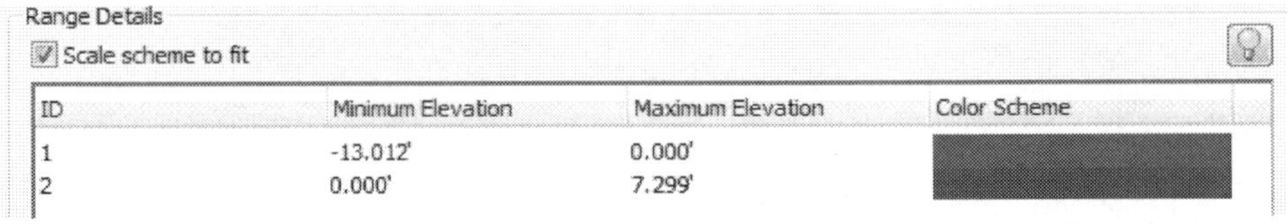

Range Details

Scale scheme to fit

ID	Minimum Elevation	Maximum Elevation	Color Scheme
1	-13.012'	0.000'	
2	0.000'	7.299'	

***Figure 4-12** Partial view of the **Range Details** area for Elevation analysis with two bands*

Creating the New Legend Style

In this section, you will create a new legend style.

1. In the **Analysis** tab, choose the down arrow button from the **Legend** area and then choose the **Create New** option from the flyout displayed, as shown in Figure 4-13. On doing so, the **Table Style - New Elevation Table Style** dialog box is displayed.

***Figure 4-13** Choosing the **Create New** option from the flyout*

2. In the **Information** tab, enter **Cut and Fill Levels** in the **Name** edit box.

3. Choose the **Data Properties** tab from the dialog box. In the **Structure** area of this tab, double-click on the table name **Elevations Table**; the **Text Component Editor - Column Contents** dialog box is displayed.

4. Delete the existing text in the **Text Editor** text box and then enter **Cut and Fill Levels** in it, as shown in Figure 4-14.

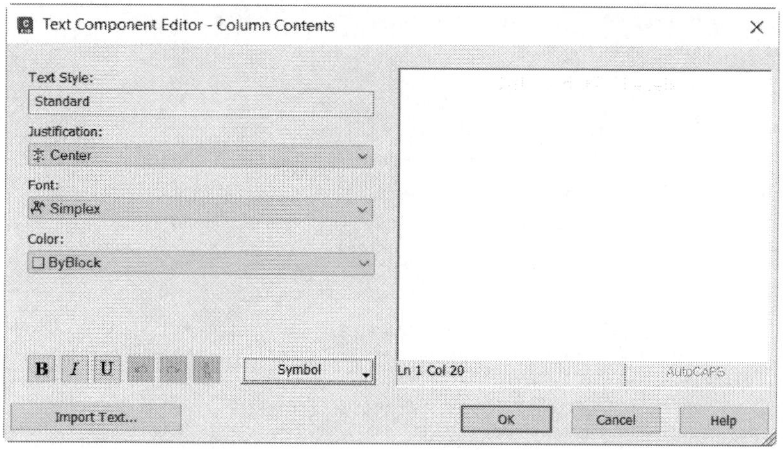

***Figure 4-14** The text entered in the **Text Editor** text box*

5. Choose the **OK** button to close the dialog box; the **Table Style - Cut and Fill Levels** dialog box is displayed. Notice that the name of the table in the **Structure** area is changed to **Cut and Fill Levels**.

6. In this dialog box, double-click on the **Minimum Elevation** header in the **Structure** area; the **Text Component Editor - Column Contents** dialog box is displayed.

7. Delete the existing text in the **Text Editor** text box and then enter **Cut Level** in it. Next, choose the **OK** button; the dialog box is closed. Notice that the **Minimum Elevation** column in the **Structure** area of the **Table Style - Cut and Fill Levels** dialog box is renamed as **Cut Level**.

8. Double-click on the **Maximum Elevation** header; the **Text Component Editor - Column Contents** dialog box is displayed.

9. Select and then delete the existing text in the **Text Editor** text box and enter **Fill Level**.

10. Next, choose the **OK** button; the **Text Component Editor - Column Contents** dialog box is closed.

11. Choose the **Add Column** button; a new column is added to the **Structure** area. 

12. Double-click on the header of the newly added column; the **Text Component Editor- Column Contents** dialog box is displayed.

13. In the **Text Editor** text box of the dialog box, enter **Net Volume** and then choose the **OK** button; the **Text Component Editor - Column Contents** dialog box is closed.

14. Now, double-click in the **Column Value** cell of the **Net Volume** column; the **Text Component Editor - Column Contents** dialog box is displayed.

15. Now, select the **Surface Range Volume** option from the **Properties** drop-down list, as shown in Figure 4-15.

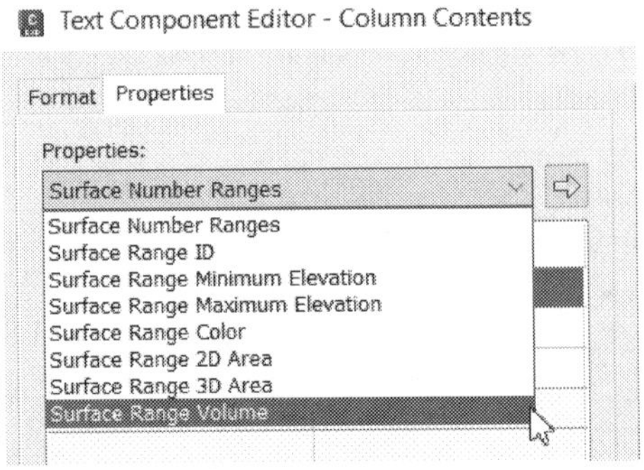

***Figure 4-15** Selecting the **Surface Range Volume** option from the **Properties** drop-down list*

16. Choose the arrow button to add this property to the **Text Editor** window. 

17. Next, choose the **OK** button; the **Text Component Editor - Column Contents** dialog box is closed.

18. Now, choose the **Display** tab from the **Table Style - Cut and Fill Levels** dialog box.

19. Set the **Overall Border** color display to **By Layer**. Also set color for the **Title Separator**, **Header Separator**, and **Data Separator** components to **blue**. Next, set the color of the **Title Text**, **Header Text**, and **Data Text** components to **red**.

20. Next, click on the bulb icon in the **Visible** column of the **Title Area Fill**, **Header Area Fill**, and **Data Area Fill** to turn off the display. Next, choose the **OK** button to close the **Table Style - Cut and Fill Levels** dialog box.

21. Select the **Preview** check box in the **Analysis** tab of the **Surface Properties - Volume Surface** dialog box to view the new **Cut and Fill Levels** legend table in the preview, as shown in Figure 4-16.

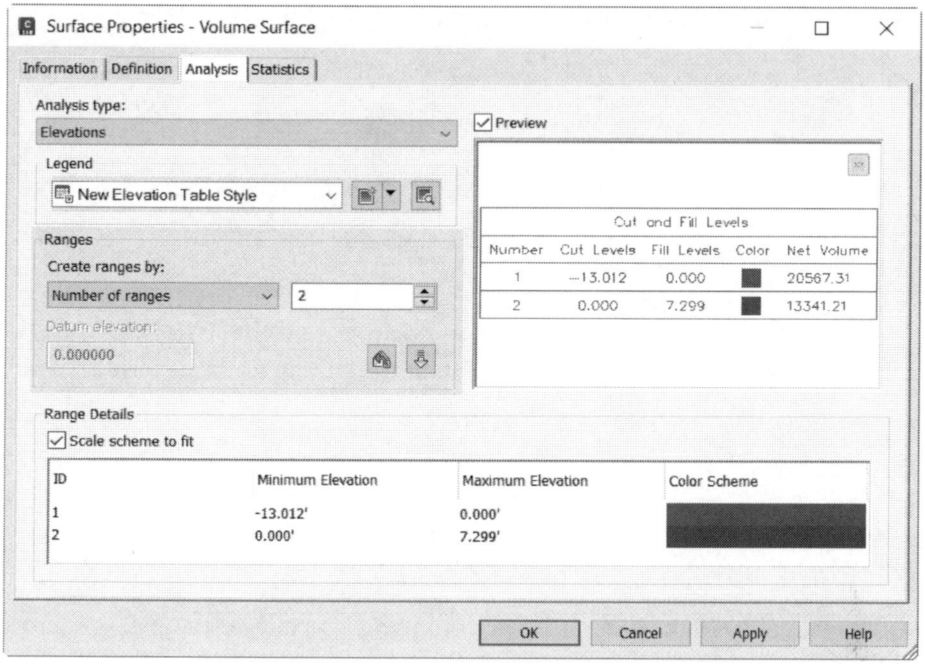

Figure 4-16 *The **Analysis** tab displaying the **Cut and Fill Levels** legend table in the **Preview** window*

22. Choose the **OK** button to close the **Surface Properties - Volume Surface** dialog box.

Adding the Legend Table to the Drawing

1. Choose the **Add Surface Legend Table** tool from **Annotate > Labels & Tables > Add Tables** drop-down; you are prompted to select the surface.

2. Press ENTER; the **Select a Surface** dialog box is displayed.

3. Select **Volume Surface** and choose **OK**; the dialog box is closed. Also, you are prompted to specify the type of the table.

4. Press ENTER to accept the default **Elevations** option; you are prompted to specify the behavior of the table.

5. Press ENTER to accept the default **Dynamic** option; you are prompted to select the upper left corner.

6. Click at the required location in the drawing; the legend table is added to the drawing and displays the cut and fill levels along with net volume, as shown in Figure 4-17.

Cut and Fill Levels				
Number	Cut Level	Fill Level	Color	Net Volume
1	−13.012	0.000	■	20567.31
2	0.000	7.299	■	13341.21

Figure 4-17 *The legend table displaying the cut and fill levels along with net volume*

Computing the Bounded Volume

1. In the **Prospector** tab of the **TOOLSPACE** palette, choose the **Volume Surface** subnode from the **Surfaces** node and right-click on it; a shortcut menu is displayed.

2. Choose the **Edit Surface Style** option from the shortcut menu; the **Surface Style - Elevation Banding (2D)** dialog box is displayed.

3. In this dialog box, choose the **Display** tab. Turn off the visibility of **Elevations** component in this tab. Choose the **OK** button to close the dialog box.

4. Enter **ReportSurfBoundedVolume** in the command line and press ENTER twice; the **Select a Surface** dialog box is displayed.

5. Choose the **Volume Surface** option from this dialog box.

6. Choose the **OK** button; the dialog box is closed and you are prompted to select the bounding polygon.

7. Select the **Finished Ground** displayed in magenta color from the drawing, as shown in Figure 4-18.

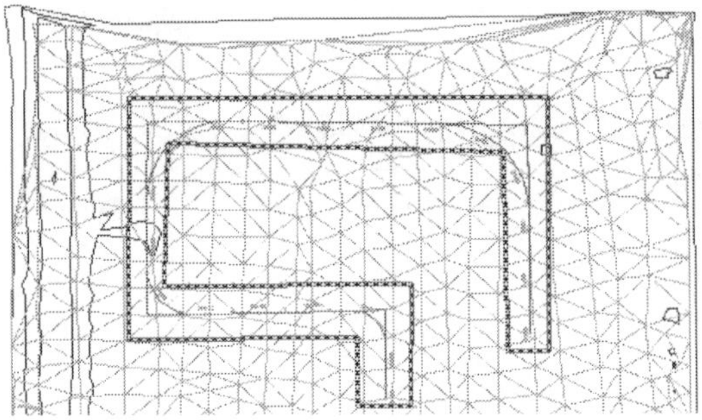

Figure 4-18 *Selecting the **Finished Ground** from the drawing*

8. Press the F2 key; the **AutoCAD Text Window** is displayed showing the cut, fill, and net volumes of the area bounded by the polygon, as shown in Figure 4-19.

```
Select a surface <or press enter key to select from list>:

Surface:  Existing Ground
Current datum elevation: 0.000'
Select bounding polygon or [Datum]: *Cancel*

Command: *Cancel*

Command: REPORTSURFBOUNDEDVOLUME


Select a surface <or press enter key to select from list>:

Surface:  Volume Surface
Current datum elevation: 0.000'
Select bounding polygon or [Datum]:
          Net volume = 35.00 Cu. Yd.<Cut>
                 Cut = 35.00 Cu. Yd.
                Fill = 0.00 Cu. Yd.
Net volume (adjusted) = 35.00 Cu. Yd.<Cut>
      Cut (adjusted) = 35.00 Cu. Yd.
     Fill (adjusted) = 0.00 Cu. Yd.
Current datum elevation: 0.000'

Select bounding polygon or [Datum]:
```

Figure 4-19 *The* ***AutoCAD Text Window*** *displaying the cut, fill, and net volumes of the bounded area*

Saving the File

1. Choose **Save As** from the **Application Menu**; the **Save Drawing As** dialog box is displayed.

2. In this dialog box, browse to the following location:

   C:\c3d_2024\c04_c3d_2024_tut

3. In the **File name** edit box, enter **c04_tut01a**.

4. Choose the **Save** button; the file is saved with the name *c04_tut01a.dwg* at the specified location.

5. Choose **Close** from the **Application Menu** to close the file.

EXERCISE

1. Log on to *www.cadsofttech.com* and browse to *Books > Civil/GIS Books > AutoCAD Civil 3D > AutoCAD Civil 3D 2024 for Novices*. Next, select *c04_c3d_2024_exer.zip* file from the **Exercise Files** drop-down list. Choose the **Download** button to download the data file.

2. Now, save and extract the downloaded folder to the following location:

   C:\c3d_2024

Note

*While opening the exercise files, the **PANORAMA** window may appear with an error message. Close this window to proceed further.*

Exercise 1

Open the *c04_c3d_2024_ex01.dwg* file and run the Contour analysis. Also, add contour labels. Figures 4-20 and 4-21 are given for your reference.

(Expected time: 30 min)

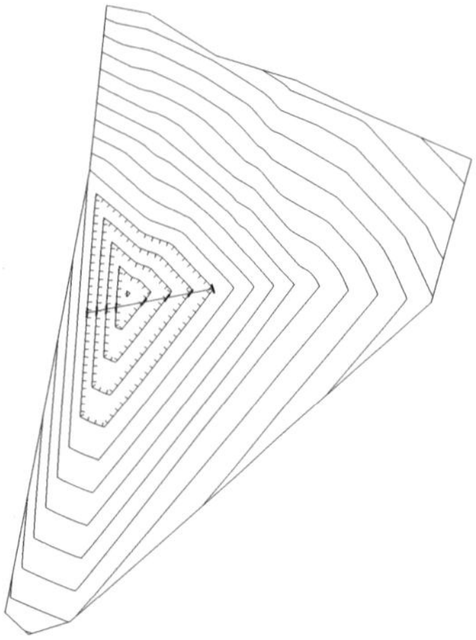

Figure 4-20 *Surface after contour analysis*

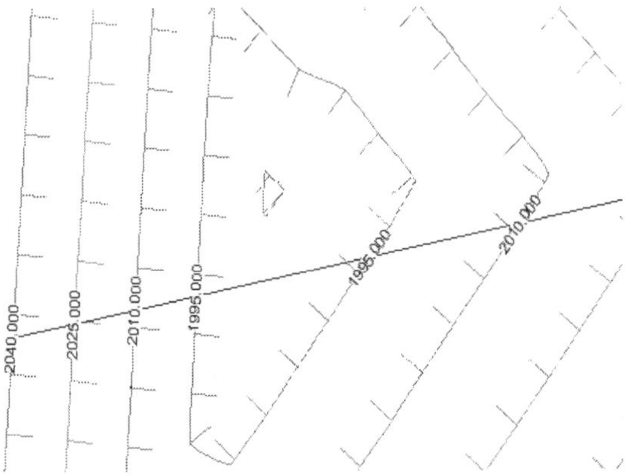

Figure 4-21 *Depression contours and contour labels*

Right-click on **Surface 1** in the **Prospector** tab and choose **Edit Surface Style** from the shortcut menu. Specify the following settings in the **Contours** tab of the **Surface Style** dialog box:

Number of Ranges: **5**
Minor Interval: **15.000'**
Display Depression Contours: **True**
Tick Mark Length: **3.000'**
Smooth Contours: **True**

Chapter 5

Alignments

Learning Objectives

After completing this chapter, you will be able to:

- *Create alignment and its components*
- *Understand methods of creating alignments*
- *Use alignment layout tools*
- *Use different methods of editing alignment*
- *Understand and create superelevation view*

ALIGNMENTS

An alignment is a linear Civil 3D object with coordinates and consists of lines, curves, and spirals. The alignment represents features such as the center line of a road, highway, tunnel, pavement edges, sidewalks, pipe lines, and so on. Alignment can represent existing or proposed features. It helps surveyors and engineers in creating the layouts of the projects easily and effectively. Defining the alignment is the first and basic step in designing a road or a pipeline. While starting up a new transportation project, the alignments should be designed carefully as they also serve as the geometries that control the layout and construction of the highway. A well designed road alignment reduces the cost of maintenance and vehicle operations. It also provides comfortable driving and road safety.

In AutoCAD Civil 3D, an alignment can exist as an independent object or as an object dependent on other Civil 3D objects such as profiles, cross-sections, and parcels. Alignments are dynamic in nature implying that the changes made by the user are dynamically reflected in all inter-related Civil 3D objects. In AutoCAD Civil 3D, the process of drafting and modifying an alignment layout is easy.

Types of Alignments

Alignments can be categorized into two types: vertical and horizontal. Horizontal alignment represents the geometry of an object in a plan view. It includes geometry of the object along straight paths and curves in a horizontal plane. Vertical alignment represents the elevation of points along the length of the object. The vertical alignment is represented by the longitudinal profile of the object and it includes the vertical curves and gradients.

The vertical and horizontal alignments should meet the required design standards so as to yield maximum efficiency. The criterion for designing alignments is discussed next.

Key Points for Designing an Alignment

Creating an alignment requires careful and thorough study of the area. The key points which should be considered while designing a road alignment are discussed next.

1. Designed alignment should be short and straight. The construction along the designed alignment should require minimum cutting and embankment. It should also avoid region with poor soil structure for a stable foundation of the structure.

2. An alignment should be safe for construction, operation, and maintenance from the stability point of view of the constructed structure. It should also be safe for the traffic operation and must have smooth curves and gradients.

3. An alignment should be designed on a higher ground so as to allow good drainage. Also, it should have a minimum number of cross drainages along its length.

4. The structure when constructed along the proposed alignment should add to the aesthetic beauty of the region.

5. Finally, the alignment should be such that the cost of construction, maintenance, and operation of the project is economized leading to a good project cost benefit ratio.

Components of Alignment

The two main components of an alignment are stations and offsets. Stations represent segments or distances into which an alignment is divided. Each alignment starts and terminates at a station. Each station is represented by a value that specifies the distance of that station from the start or the first station of the alignment. For example, the value 5+40 denotes that the station is located at 540' from the start of the alignment. The specification of the first station in an alignment depends on its length or the project requirement. The values of the stations automatically increase toward the last station of the alignment.

Offsets, the other important components of an alignment, represent the location of the existing or proposed objects in the site. These offsets help the surveyors to locate the nearby objects by taking the alignment as the reference or the base line. If we take the direction of the increasing stations as the reference then the offsets on the right of the alignment have a positive value and the offsets on the left have a negative value. For example, a tree with a station 2+50 and an offset -16.3 indicates that it is located at a distance of 250' from the first station and at a distance of 16.3' to the left of the alignment.

CREATING ALIGNMENTS

You can create alignments from polylines or by using the **Alignment Layout** tools. You can also create an alignment from the pipe network or by importing the data from landXML files. The basic information required to create an alignment are the name, type, and the style of the alignment. Civil 3D also allows you to use the criteria-based designs for creating alignments.

Once an alignment is created, it is added to the **Alignment** node in the **Prospector** tab of the **TOOLSPACE** palette.

Creating Alignments Using Polylines

Ribbon:	Home > Create Design > Alignment drop-down > Create Alignment from Objects
Command:	CREATEALIGNMENTENTITIES

The simplest and the most common method of creating an alignment is by using polylines. Using this method, you can automatically add curves between the tangents in the alignments. You can create an alignment from a polyline in two steps. The first step is to sketch the polyline and the second step is to convert the polyline into an alignment. To sketch the polyline, invoke the **Polyline** tool from the **Draw** panel of the **Home** tab; you will be prompted to specify the start point of the polyline. Click in the drawing area to specify the start point of the polyline. Click again when prompted to specify the next point. You can continue clicking to specify the points or press ENTER to terminate the command. On exiting the command, a polyline will be created. Now to convert this polyline into an alignment, choose the **Create Alignment from Objects** tool from the **Create Design** panel; you will be prompted to select a line, an arc, or a polyline object. In the drawing, select the polyline that you have drawn. Also, you can enter **X** in the command line to select the polyline from within the Xref drawing. On selecting the polyline from the drawing or from the Xref drawing, you will be prompted to press ENTER to accept the alignment direction or to enter **R** to reverse the direction. Press ENTER to accept the default alignment direction; the **Create Alignment from**

Objects dialog box will be displayed, as shown in Figure 5-1. This dialog box is used to specify the information about the alignment.

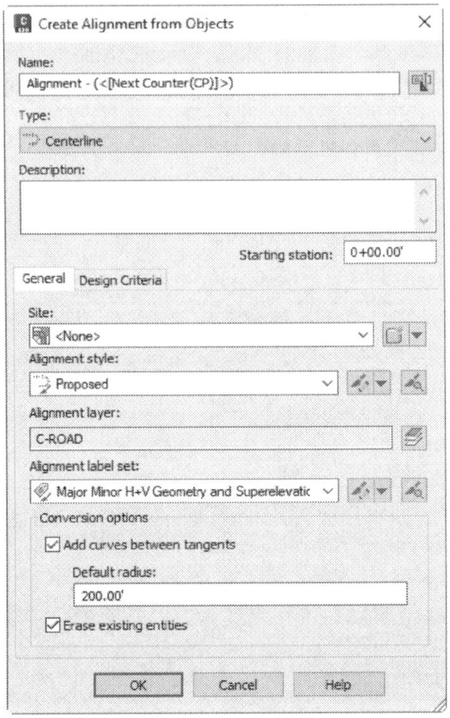

Figure 5-1 *The* ***Create Alignment from Objects*** *dialog box*

Creating Alignments Using the Alignment Layout Tools

Ribbon: Home > Create Design > Alignment drop-down > Alignment Creation Tools
Command: CREATEALIGNMENTLAYOUT

AutoCAD Civil 3D allows you to create an alignment based on some intricate designs by using the alignment layout tools. These tools can be accessed from the **Alignment Layout Tools** toolbar. These tools enable you to incorporate your engineering designs in the alignment and create an alignment according to the design criteria. For example, if you want the alignment curve to pass through a particular point, you can create the layout of the alignment accordingly by using the required curve tool from the toolbar. These sophisticated layout tools help you create an alignment layout using the curves, spirals, points of intersection (PIs), and so on.

Creating an alignment using the alignment layout tools is a two-step process. The first step includes naming the alignment, specifying the site on which the alignment will be created, selecting the alignment style, and the alignment label style. The second step involves the use of the alignment layout tools to draft the alignment.

To create an alignment using the layout tools, choose **Alignment Creation Tools** from the **Create Design** panel; the **Create Alignment - Layout** dialog box will be displayed, as shown in

Figure 5-2. The options in this dialog box are the same as those discussed earlier for the **Create Alignment from Objects** dialog box.

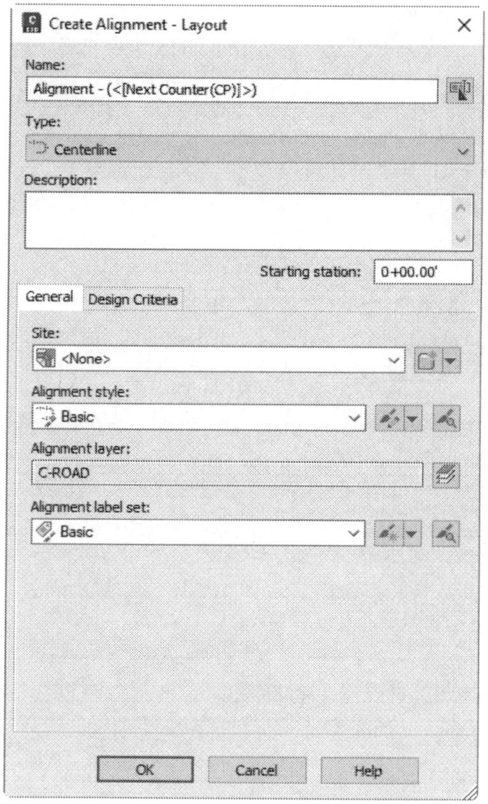

Figure 5-2 *The* ***Create Alignment - Layout*** *dialog box*

ALIGNMENT LAYOUT TOOLS TOOLBAR

The tools in the **Alignment Layout Tools** toolbar are used to design an alignment using various constraints explained earlier. You can invoke this toolbar to edit the alignment geometry. To do so, select the alignment from the drawing and then right-click; a shortcut menu will be displayed. In this shortcut menu, choose the **Edit Alignment Geometry** option; the **Alignment Layout Tools - <alignment name>** toolbar will be displayed. The tools available in this toolbar are discussed next. Figure 5-3 shows the **Alignment Layout Tools - <alignment name>** toolbar.

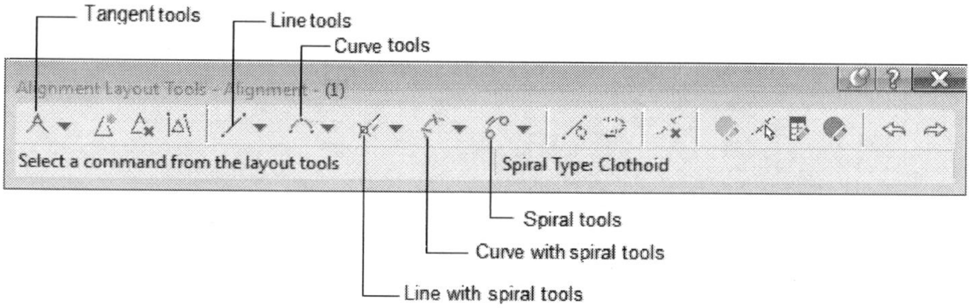

Figure 5-3 *The* ***Alignment Layout Tools - Alignment - (1)*** *toolbar*

CREATING ALIGNMENTS USING THE BEST FIT METHOD

The best fit method is used to get the best fitting alignment using the various Civil 3D objects such as the COGO points, AutoCAD points, lines, and arcs. You can access the tools for creating the best fit entities from the **Best Fit** drop-down in the **Draw** panel of the **Home** tab. You can use these tools to easily complete the complex alignment designs. Besides polylines, Civil 3D also allows you to convert lines and curves into an alignment.

The tools in the **Best Fit** drop-down help you create lines, arcs, and parabolas from various AutoCAD entities. You can also create a best fit alignment by picking points from the screen.

These tools are especially used for the rehabilitation projects where you need to resurvey the existing road. In these type of projects, the existing survey data may not match with the original survey data. This situation demands that you create an alignment from the available data. You can use the tools from the **Best Fit** drop-down to edit the lines and curves to match the actual survey data in the best possible way. These tools can also be used for both alignments and profiles. To access these tools, click on the down arrow next to the **Best Fit** drop-down in the **Draw** panel. On doing so, a list of tools will be displayed, as shown in Figure 5-4. Invoke the required tool from the drop-down. These tools are discussed next.

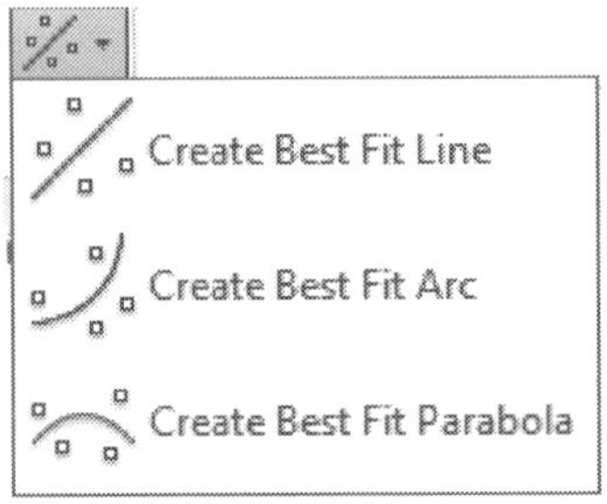

***Figure 5-4** The tools in the **Best Fit** drop-down*

Creating Lines, Arcs, and Points as Entities by Using the Best Fit Entities Tool

Ribbon: Home > Draw > Best Fit drop-down > Create Best Fit Line / Create Best Fit Arc / Create Best Fit Parabola
Command: CREATELINEBYBESTFIT / CREATEARCBYBESTFIT / CREATEPARABOLABYBESTFIT

To create a line of best fit, choose the **Create Best Fit Line** tool from the **Draw** panel; the **Line by Best Fit** dialog box will be displayed, as shown in Figure 5-5. You can create the best fit entity by selecting one of the four different options available in this dialog box.

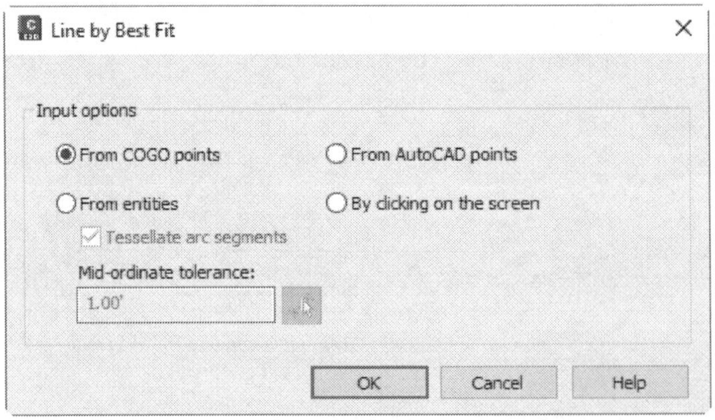

***Figure 5-5** The **Line by Best Fit** dialog box*

To create a line using the Civil 3D points, select the **From COGO points** radio button from the **Line by Best Fit** dialog box and then choose the **OK** button; the **Line by Best Fit** dialog box will close and you will be prompted to select the point objects. Select the point objects in the drawing and press ENTER; the **PANORAMA** window with the **Regression Data** will be displayed. You can edit the regression data in the **PANORAMA** window as required and choose the **Create Entity** button to create a line of best fit.

To create a line from the middle of an entity, select the **From entities** radio button; the **Mid-ordinate tolerance** edit box will be enabled in which you can specify the mid-ordinate distance.

Similarly, you can create the best fit arc and parabola entities by choosing the **Create Arc/Create Parabola** tool from the **Draw** panel. The steps involved in creating these entities are the same as those explained for creating the line entity.

To convert the **Best Fit** entities into an alignment, choose the **Convert AutoCAD line and arc** tool from the **Alignment Layout Tools** toolbar and select the required line and arc to convert them into an alignment.

CRITERIA-BASED ALIGNMENT DESIGN

In AutoCAD Civil 3D, the criteria-based design feature ensures that the design meets the required project standards. Before you understand the workflow of creating the criteria-based alignment design, it is important to understand various design criteria.

Design Criteria

The design criteria file is an XML file within AutoCAD Civil 3D that consists of the common and minimum design criteria for creating alignments and profiles. The design criteria values are specified for design speeds, minimum speed, superelevation, curve radius, and length of the individual alignment entities. There are two design criteria files available in Civil 3D, one in **Metric Units** and other in **Imperial Units**. These design criteria are based on AASHTO (American Association of State and Highway Transport Officials) standards. The design criteria files can be customized to match the required standards for the project and then saved on a local server. A design file on a server can be accessible to all members of the team.

Creating Connected Alignments

Ribbon:	Home > Create Design > Alignment drop-down > Create Connected Alignment
Command:	CREATEALIGNMENTCONNECTED

The connected alignment capabilities in AutoCAD Civil 3D allow you to create a new dynamically linked alignment and profile that enables transitions between two intersecting alignments and their profiles. You can use this feature to create a curb return, an exit ramp, a merging/diverging road, or to connect an existing road with a proposed road.

Choose the **Create Connected Alignment** tool from the **Alignment** drop-down in the **Create Design** panel of the **Home** tab; first you will be prompted to select the first alignment and then the second alignment from the drawing. Now, you will be prompted to pick a location for connected alignment and then press ENTER to accept the location; the **Create Connected Alignment** dialog box will be displayed, as shown in Figure 5-6.

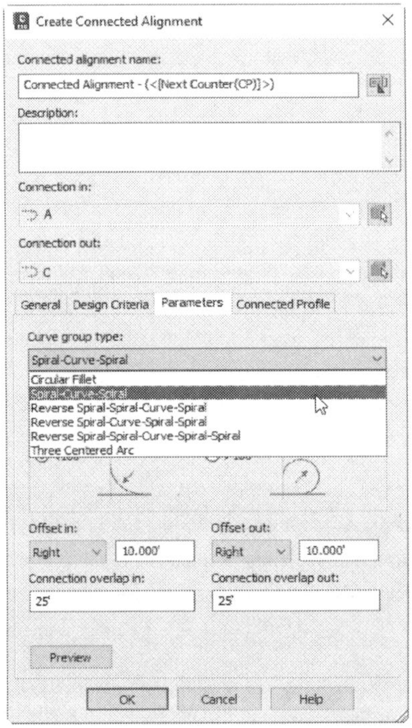

Figure 5-6 *The* ***Create Connected Alignment*** *dialog box*

Design Checks

Design checks are created to check the criteria which are not included in the in-built design criteria files. These checks are basically mathematical expressions defined to verify that the alignment entities such as lines, curves, and spirals comply with the criteria set for the alignment. You can create individual design checks for lines, spirals, curves, and tangents and save them in a design check set that can be applied to an alignment. As explained earlier, a design check set is a collection of different design checks. You can create a new design check set or edit an existing one by choosing **Alignment > Design Checks > Design Check Sets** in the **Settings** tab in the **TOOLSPACE** palette.

Working with Design Checks in a Design Check Set

To add a design check in a design check set, choose the **Settings** tab and expand the **Alignment** node. Now, choose the **Design Check Sets** subnode to view the default name of the design check set. Next, right-click on the **Design Check Sets** and choose **New** from the shortcut menu displayed; the **Alignment Design Check Set - New Alignment Check Set** dialog box will be displayed. The **Information** tab will be chosen by default in this dialog box. In this tab, enter a name of the check set in the **Name** edit box. Next, choose the **Design Checks** tab. In this tab, select the required option from the **Type** and **<Geometry Type> Checks** drop-down lists. Next, choose the **Add** button to add the selected design check and type in the respective columns, refer to Figure 5-7. Then, choose the **OK** button to close the dialog box.

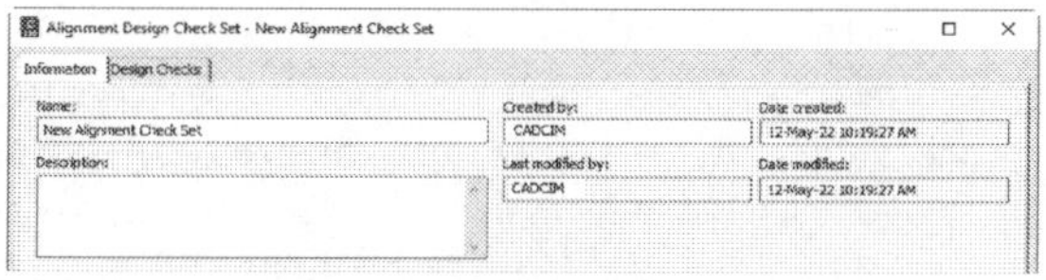

***Figure 5-7** The partial view of the **Alignment Design Check Set - New Alignment Check Set** dialog box showing some design checks*

Creating a Design-based Alignment

Ribbon: Home > Create Design > Alignment drop-down > Alignment Creation Tools
Command: CREATEALIGNMENTLAYOUT

To create a design-based alignment, choose the **Alignment Creation Tools** tool in the **Create Design** panel from the **Home** tab; the **Create Alignment - Layout** dialog box will be displayed. Choose the **Design Criteria** tab in this dialog box. The **Design Criteria** tab is used to specify the options for creating the criteria-based alignment.

Editing the Design Criteria

You can edit and customize the design criteria according to your required standards. To edit design criteria, select the required alignment in the current drawing; the **Alignment:<Alignment name>** tab will be displayed. In this tab, choose the **Design Criteria Editor** tool from the **Modify** panel; the **Design Criteria Editor - <design criteria file name.xml>** dialog box will be displayed, as shown in Figure 5-8.

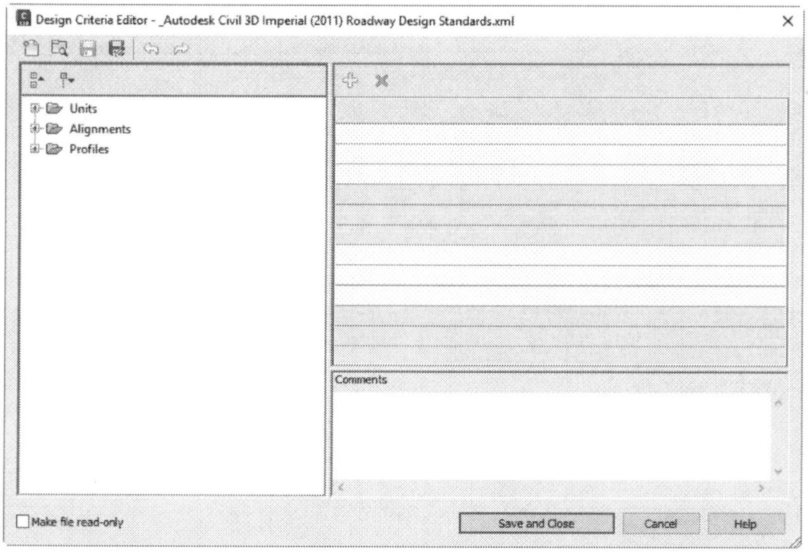

***Figure 5-8** The **Design Criteria Editor - _Autodesk Civil 3D Imperial (2011) Roadway Design** dialog box*

On the left pane of this dialog box, three collections are displayed: **Units**, **Alignments**, and **Profiles**. Expand the **Alignments** collection; four sub-collection will be displayed namely the minimum radius tables, superelevation attainment methods, superelevation tables, and widening methods. You can expand the required sub-collection for which you want to edit the design criteria. For example, to change the minimum radius in the design, expand the **Minimum Radius Table** sub-collection and select any of the tables listed in it. On doing so, the values of different design speeds and the minimum radii will be displayed in columns on the right pane of the dialog box. Click on the required radius value and change it at the specified design speed according to the design criteria.

Similarly, expand other sub-collections. Select them to display the design parameters in columns on the right pane of the dialog box. You can edit the values as required. You can also modify the units for the design criteria file by using the **Units** collection. The **Profiles** collection allows you to edit the profile criteria if the alignment consists of any profiles.

ALIGNMENT PROPERTIES

Ribbon:	Alignment > Modify > Alignment Properties drop-down > Alignment Properties
Command:	EDITALIGNMENTPROPERTIES

To view and edit the alignment properties, select the required alignment from the drawing; the **Alignment: <Alignment Name>** tab will be displayed. Next, choose the **Alignment Properties** tool from the **Modify** panel; the **Alignment Properties - <alignment name>** dialog box will be displayed. This dialog box is used to assign new name and object style, and set station equations, design criteria, and superelevation properties to an alignment. You can also view the profile information if the alignment consists of profiles. You will learn about the profiles in the next chapter. The tabs in this dialog box are discussed next.

Tip

You can also select the required alignment name from the ***Alignments*** *collection in the* ***Prospector*** *tab of the* ***Toolspace*** *palette and right-click to display the shortcut menu. Choose* ***Properties*** *from the shortcut menu to display the* ***Alignment Properties*** *dialog box.*

Information Tab

The **Information** tab displays the name and object style of the selected alignment. You can modify the name by using the **Name** edit box. The style can be changed by selecting an option from the **Object style** drop-down list.

Station Control Tab

This tab help you to change the station reference point and add station equations along an alignment.

Design Criteria Tab

This tab is used to assign or edit the design criteria and speeds. This tab can also be used to assign variable speed limits for stations along an alignment. However, only one design speed area can be assigned to any given station.

ALIGNMENT STYLES

Alignment styles help you control the appearance of the objects, labels, and tables of an alignment. AutoCAD Civil 3D consists of various in-built styles that are applied automatically to alignment objects. You can also create new alignment styles and control the display of alignment objects, alignment labels, and tables as required. The alignment styles are created and managed in the **Settings** tab of the **TOOLSPACE** palette.

Creating an Alignment Style

The alignment styles control the visual display of the alignment entities such as lines, arcs, curves, and tangents. To create alignment styles, choose the **Settings** tab of the **TOOLSPACE** palette and expand the **Alignment** node. Right-click on the **Alignment Styles** option and choose **New** from the shortcut menu displayed; the **Alignment Style - New Alignment Style** dialog box will be displayed, as shown in Figure 5-9.

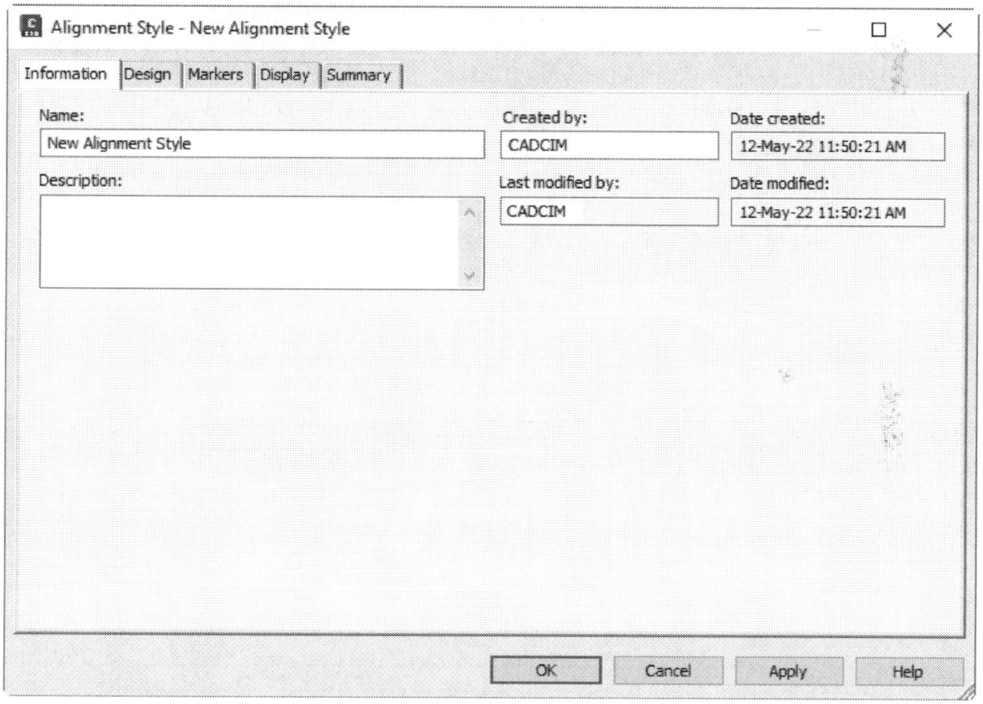

*Figure 5-9 The **Alignment Style - New Alignment Style** dialog box*

Editing the Current Alignment Style

Ribbon:	Alignment > Modify > Alignment Properties drop-down > Edit Alignment Style
Command:	EDITALIGNMENTSTYLE

To edit an existing style, select the alignment from the drawing; the **Alignment:<Alignment Name>** tab will be displayed. Next, choose the **Edit Alignment Style** tool from the **Modify** panel; the **Alignment Style** dialog box with the alignment style name will be displayed. Edit the alignment style as required and choose the **OK** button to close the dialog box.

Tip
*You can also edit the alignment style using the **Settings** tab. To do so, choose the **Settings** tab and expand the **Alignments** node. In this node, expand the **Alignment Styles** sub node. Next, right-click on the existing alignment style name and choose **Edit** from the shortcut menu displayed.*

ALIGNMENT LABEL STYLES

The alignment style is created to control the display of the alignment objects whereas the alignment label styles are created to control the display of text and labels of these alignment objects. The alignment labels are automatically generated in Civil 3D along the alignment at some specified points such as start and end points of the alignment, major and minor stations, at the point of intersections, and so on. The display of the alignments is controlled by the alignment styles. Civil 3D has a number of in-built label styles to label different components of an alignment such as stations, station offsets, lines, curves, spirals, points of intersections, and so on. You can manage alignment label style by specifying the required alignment label set. A label set is a group of label styles that are defined for various alignment components. The label set is created by using the options in the **Settings** tab of the **TOOLSPACE** palette.

Creating an Alignment Label Style

To create a label style, expand the **Alignment** collection and then expand the **Label Styles** node. You will see that the **Label Styles** node consists of different subnodes of various alignment objects such as stations, station offsets, lines and curves. Each of these alignment objects has varied label styles.

The first step to create an alignment label style is to define a label style for each of these elements and then add these label styles to the label set. The label set is a combination of different label styles that you want to apply to the alignment. It helps you easily assign the labels to the alignment.

To create a label style, expand the **Alignment** node from the **Settings** tab and then expand the **Label Styles** node; the different subnodes will be displayed. Expand the required subnode for which you want to define the label style. For example, to define a label style for the stations and the station offsets of the alignment, expand the **Station** and **Station Offsets** subnodes; the **Station** subnode consists of different elements such as the major and minor stations, geometry points and the design speeds for which you need to define the label styles. You can define labels for all the elements or the required elements. For example, to define a label style for the major station, select and right-click on **Major Station** and then choose **New** from the shortcut menu displayed; the **Label Style Composer - New Major Station Label Style** dialog box will be displayed, as shown in Figure 5-10.

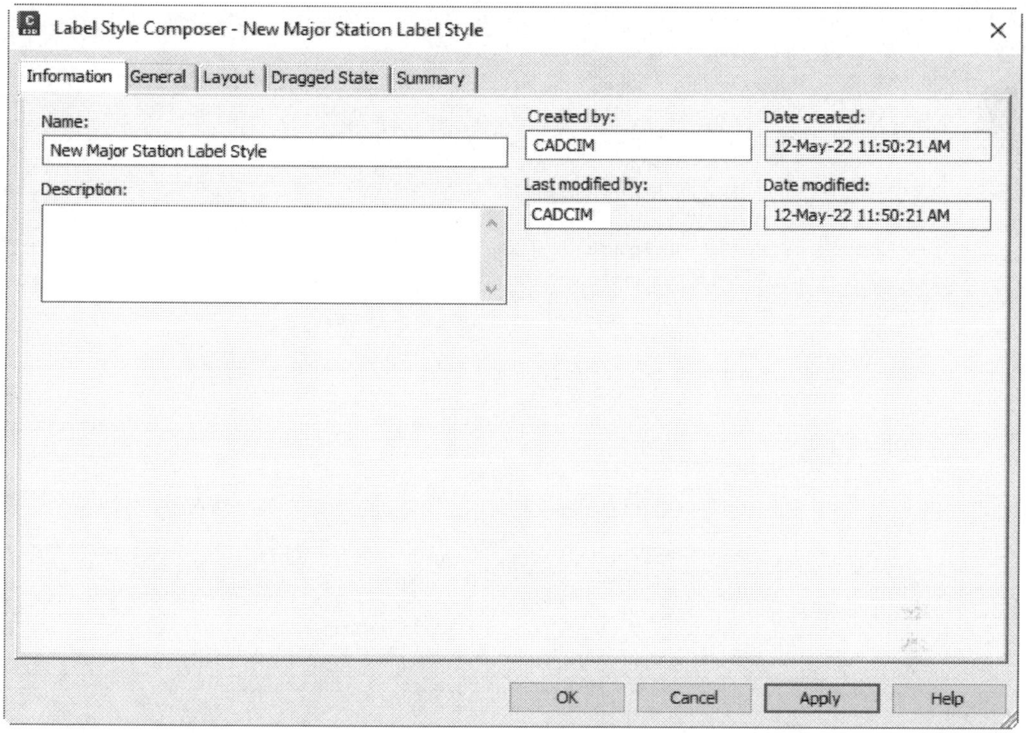

Figure 5-10 The ***Label Style Composer - New Major Station Label Style*** *dialog box*

Creating a Label Set

To create a label set, expand the **Alignment** node from the **Settings** tab. Next, expand the **Label Styles** node and right-click on the **Label Sets**; a shortcut menu will be displayed. Choose **New** from the shortcut menu; the **Alignment Label Set - New Alignment Label Set** dialog box will be displayed.

Adding Alignment Labels

Ribbon: Annotate > Labels & Tables > Add Labels drop-down > Alignment > Add Alignment Labels
Command: ADDALIGNLABELS

Once you have created and configured your own styles for the alignment objects, you can add them in the drawing. To do so, choose the **Add Alignment Labels** from the **Labels & Table** panel; the **Add Labels** dialog box will be displayed, as shown in Figure 5-11.

Select the object to be labeled from the **Feature** drop-down list. Next, select the required label type and choose the **Add** button; you will be prompted to select the object to be labeled. You can label a single alignment segment or all segments together by selecting the **Single Segment** or **Multiple Segment** option from the **Label type** drop-down list. Similarly, you can select the **Station Offset**, **Station Offset-Fixed Point**, and other options to add labels to offset stations and tangent intersection points in the alignment. After selecting the label type, choose the **Add** button and follow the prompts to label the alignment segments. Next, choose the **Close** button to close the **Add Labels** dialog box.

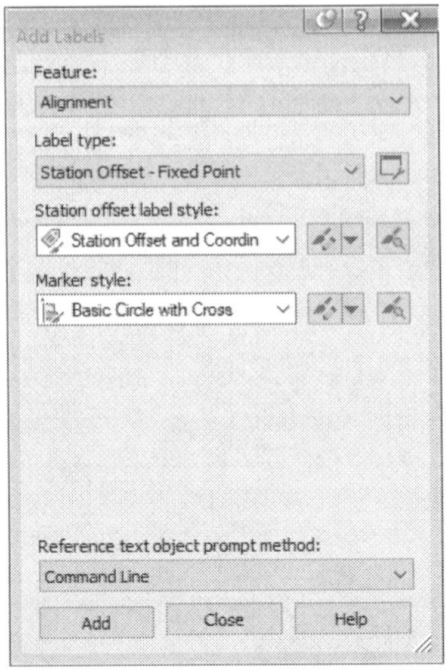

Figure 5-11 *The **Add Labels** dialog box*

EDITING ALIGNMENTS

You can edit an alignment using grip editing, the **PANORAMA** window, and the **Alignment Layout Tools** toolbar. Different methods of editing an alignment are discussed next.

Editing an Alignment Layout Using the Geometry Editor

Ribbon:	Alignment > Modify > Geometry Editor
Command:	EDITALIGNMENT

You can edit the alignment geometry by using different tools from the **Alignment Layout Tools** toolbar. You can do so by adding curves, spirals, tangents, and other sub-entities to the alignment. To edit the alignment geometry, select the required alignment from the drawing; the **Alignment:<alignment name>** tab will be displayed. Choose the **Geometry Editor** tool from the **Modify** panel; the **Alignment Layout Tools - <alignment name>** toolbar will be displayed.

Editing the Alignment Using the Grip Edits

The simplest way of editing the alignment is by using the AutoCAD tools such as **Rotate**, **Move**, **Stretch**, and **Grip edits**. Every alignment has some grips that are displayed when you select it. The grips are displayed at the point of intersections, start points, end points, and center points of the curves and lines. To graphically edit the alignment, click on the grips and drag them to the required position; the alignment will be updated and redrawn.

ADDING ALIGNMENT TABLES

Ribbon: Annotate > Labels & Tables > Add Tables drop-down > Alignment

The alignment tables help you display the information about the alignment components such as the lines, curve, spiral, and segment of the alignment. The table for each component helps you to display the information in an organized way. You can create your own table styles and add the tables in the drawing.

Note

Before you add tables in the drawing, ensure that all alignment entities are labeled as you cannot create tables without labeling the alignment entities.

To add alignment tables for a line, choose the **Alignment** option from the **Add Tables** drop-down, as shown in Figure 5-12; a flyout will be displayed. Choose the **Add Line** tool from the flyout; the **Table Creation** dialog box will be displayed, as shown in Figure 5-13.

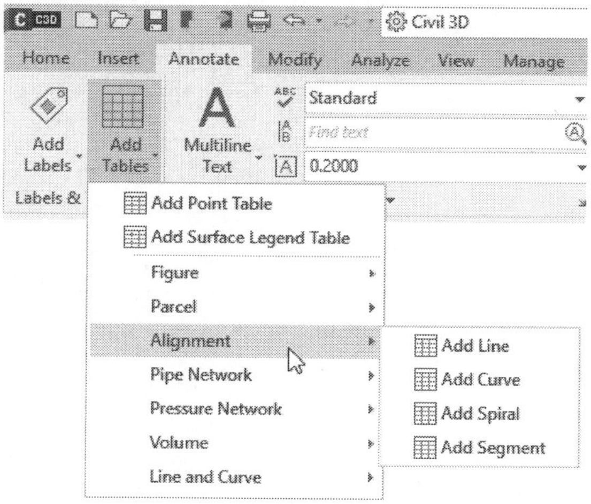

Figure 5-12 *Choosing the* ***Alignment*** *option from the* ***Add Tables*** *drop-down*

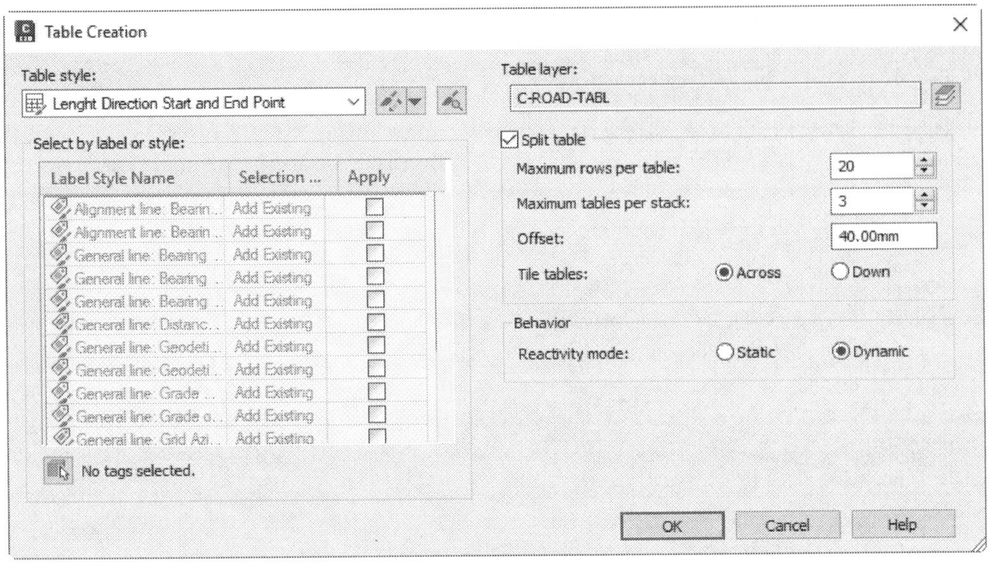

***Figure 5-13** The **Table Creation** dialog box*

In the **Table Creation** dialog box, the **Table style** drop-down list displays the default table style for the line component. You can also create a new table style by choosing the down arrow on the right of the **Table Style** drop-down list to display a flyout. Choose **Create New** from the flyout; the **Table Style - New Alignment Line Table Style** dialog box will be displayed. You can use this dialog box to create your own table style and specify the data to be displayed. The table style name will be added in the **Table Style** drop-down list. The **Table layer** text box displays the layer on which the table will be created. To change the layer of the table to be created, choose the button on the right of this text box; the **Object Layer** dialog box will be displayed. Now, select a new layer from this dialog box.

SUPERELEVATION

Superelevation is the tilting of the road surface, achieved by raising the outer edge of the pavement over the inner edge. Superelevation helps in countering the centrifugal force that can cause the vehicle to overturn or skid when moving along a curve. As mentioned earlier in this chapter, the AASHTO standards have been included in Civil 3D as the XML design criteria files. These design files contain the minimum superelevation, curve radii, design speeds, and other required values under different tables. The amount of superelevation is inversely proportional to the radius of the curve. This means that the shorter the radius of the curve, the greater will be the amount of superelevation required to maintain the design speeds. Superelevation rate is expressed in percent.

Analyzing Superelevation

In Civil 3D, you can analyze superelevation for a given alignment. You can calculate the superelevation of the alignment using the **Calculate/Edit Superelevation** tool. To modify the superelevation of the alignment, you can use the **View Tabular Editor** tool. The **Superelevation Curve Manager** can be used to specify the design criteria and edit the same for the recalculation of the superelevation.

Creating Superelevation View

Ribbon:	Alignment > Modify > Superelevation drop-down > Create Superelevation View
Command:	CREATESUPERELEVATIONVIEW

The **Create Superelevation View** tool is used to display the superelevation graphically. Choose this tool from the **Modify** panel and select the required alignment; the **Create Superelevation View** dialog box will be displayed, as shown in Figure 5-14. In this dialog box, click on the button next to the **Superelevation view name** edit box; the **Name Template** dialog box will be displayed. Enter a name in the **Name** edit box and choose the **OK** button to close the dialog box. Next, select an alignment from the **Alignment** drop-down list. Alternatively, you can select the alignment by using the **Select from the drawing** button. To do so, choose the **Select from the drawing** button on the right of the drop-down list; you will be prompted to select an alignment from the drawing. Select the alignment from the drawing; the **Create Superelevation View** dialog box will be displayed. To specify the superelevation layer, choose the button next to the **Superelevation view layer** text box. Next, to specify the style for the superelevation, select an option from the **Superelevation view style** drop-down list.

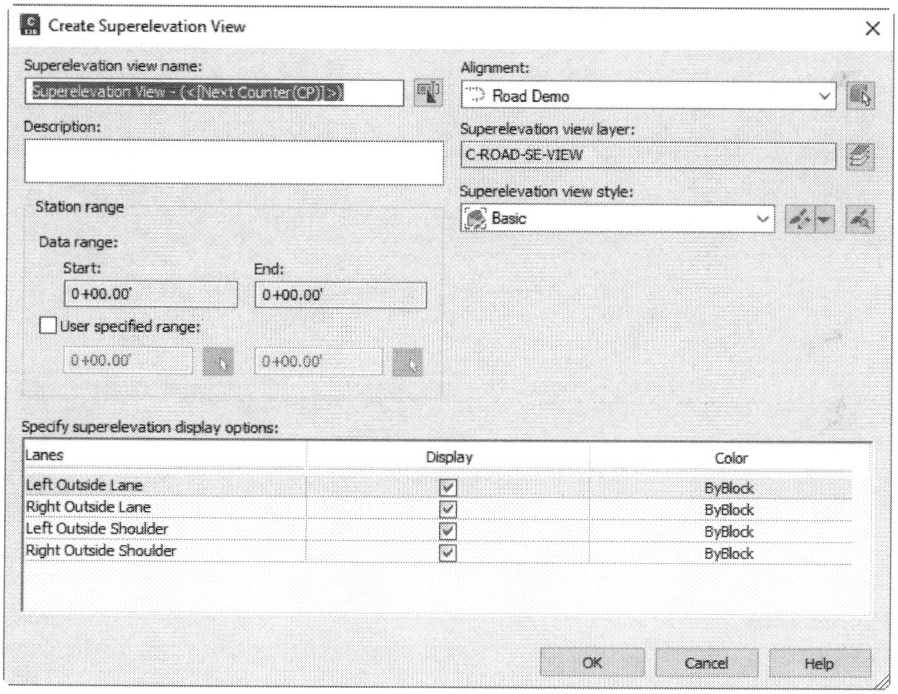

***Figure 5-14** The **Create Superelevation View** dialog box*

TUTORIAL

1. Download the *c05_c3d_2024_tut.zip* file from *https://www.cadsofttech.com,* as explained in Chapter 2.

2. Now, save and extract the downloaded file at the following location:

   c:\c3d_2024

Note
*While opening the tutorial file, the **PANORAMA** window with an error message may appear. Close this window to proceed further.*

Tutorial 1 Alignment from Polyline

In this tutorial, you will create an alignment from a polyline, as shown in Figure 5-15. You will also create and apply a new alignment style. **(Expected time: 30 min)**

The following steps are required to complete this tutorial:

a. Open the file.
b. Create alignment from a polyline.
c. Create a new alignment style.
d. Apply the new style to the alignment.
e. Save the file.

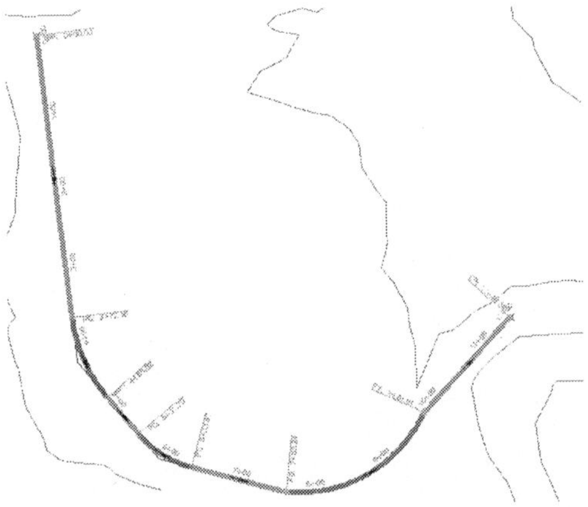

Figure 5-15 *The **Street A** alignment*

Opening the File

1. Choose **Open** from the **Application Menu**; the **Select File** dialog box is displayed.

2. In the dialog box, browse to the location *C:\c3d_2024\c05_c3d_2024_tut* where you have saved the tutorial file.

3. Select the file *c05_c3d_2024_tut01.dwg* and then choose the **Open** button to open the file. Ignore any error message, if displayed.

Creating an Alignment Polyline

1. To create an alignment from a polyline, choose the **Create Alignment from Objects** tool from **Home > Create Design > Alignment** drop-down; you are prompted to select the first line/arc/polyline or [Xref].

2. Select the red color polyline in the drawing and then press ENTER; you are prompted to specify the alignment direction.

3. Press ENTER again to select the default alignment direction; the **Create Alignment from Objects** dialog box is displayed.

4. Enter **Street A** as the alignment name in the **Name** edit box.

5. Select **Layout** from the **Alignment style** drop-down list.

6. Choose the **OK** button; the polyline is converted into an alignment with all geometry points and stations, as shown in Figure 5-16.

Figure 5-16 *Alignment of* ***Street A*** *created from polyline*

Creating a New Alignment Style

1. Choose **Alignment > Alignment Styles** from the **Settings** tab of the **TOOLSPACE** palette. You will notice a triangular symbol on the left of the **Layout** option indicating the current alignment style.

2. Select **Alignment Styles** and right-click on it; a shortcut menu is displayed.

3. Choose **New** from the shortcut menu; the **Alignment Style - New Alignment Style** dialog box is displayed.

4. Choose the **Markers** tab; various options in this tab are displayed.

5. Double-click on the button in the **Marker Style** column corresponding to the **Begin of Alignment** option; the **Pick Marker Style** dialog box is displayed. 

6. Select the **Basic X** option from the drop-down list in this dialog box, as shown in Figure 5-17. Choose the **OK** button.

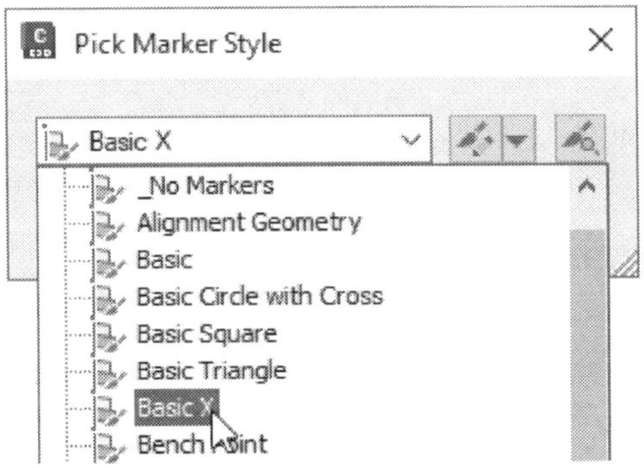

***Figure 5-17** Selecting the **Basic X** option from the drop-down list*

7. Repeat the procedure followed in steps 5 and 6 and make **Basic X** as the marker style for the **End of Alignment** option.

8. Next, choose the **Display** tab. Change the color for the **Line** component to **green**, **Curve** to **magenta,** and **Line Extensions** to **blue** in the **Color** column.

9. Change the lineweight of **Line**, **Curve**, and **Line Extensions** to **0.5** mm in the **Lineweight** column.

10. Choose the **OK** button to close the **Alignment Style - New Alignment Style** dialog box; the **New Alignment Style** is added to the **Alignment Styles** node in the **Settings** tab.

Applying the Alignment Style

1. Select the **Street A** alignment from the drawing and right-click; a shortcut menu is displayed.

2. Choose **Alignment Properties** from the shortcut menu; the **Alignment Properties - Street A** dialog box is displayed.

> **Tip**
> *You can also display the **Alignment Properties - Street A** dialog box from the **Prospector** tab of the **Toolspace** palette. To do so, expand the **Alignments** node and the **Centerline Alignments** sub node. Then, right-click on **Street A** node to display a shortcut menu. Choose **Properties** from the shortcut menu to display the **Alignment Properties - Street A** dialog box.*

3. In the **Information** tab, select the **New Alignment Style** option from the **Object style** drop-down list.

4. Choose the **Apply** button to apply the style to the alignment.

5. Then, choose the **OK** button; the **Alignment Properties - Street A** dialog box is closed and the alignment style is assigned to the **Street A** alignment.

6. Now, choose the **Show/Hide Lineweight** button from the status bar at the bottom right of the screen; the alignment entities are displayed with the assigned lineweight (0.5mm), as shown in Figure 5-18.

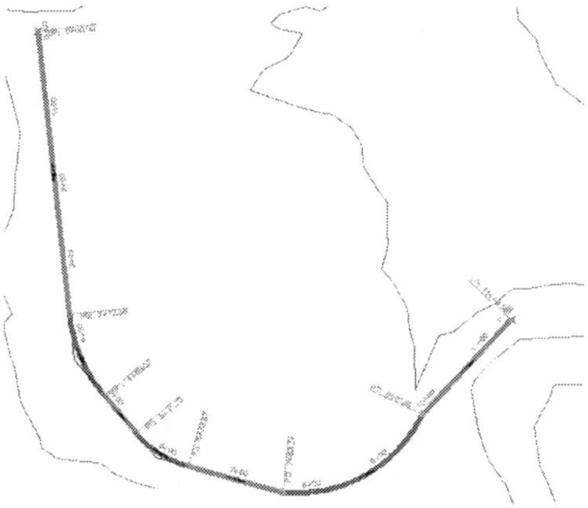

Figure 5-18 *The* ***Street A*** *alignment after assigning the New Alignment Style to it*

Saving the File

1. Choose **Save As** from **Application Menu**; the **Save Drawing As** dialog box is displayed.

2. In this dialog box, browse to the following location:

   C:\c3d _2024\c05_c3d_2024_tut

3. In the **File name** edit box, enter **c05_tut01a**.

4. Choose the **Save** button; the file is saved with the name *c05_tut01a.dwg* at the specified location.

EXERCISE

1. Log on to *www.cadsofttech.com* and browse to *Books > Civil/GIS Books > AutoCAD Civil 3D > AutoCAD Civil 3D 2024 for Novices*. Next, select *c05_c3d_2024_exer.zip* file from the **Exercise Files** drop-down list. Choose the **Download** button to download the data file.

2. Now, save and extract the downloaded folder to the following location:

   C:\c3d_2024

Note

While opening the exercise files, the ***PANORAMA*** *window may appear with an error message. Close this window to proceed further.*

Exercise 1

Open the *c05_c3d_2024_ex01.dwg* file and create an alignment using the **Create Alignment from Objects** tool. Also, calculate the superelevation values at the superelevation rate of 10%. Use the following parameters: **(Expected time: 30 min)**

1. Alignment name: **Lane 1**
2. Alignment style: **Basic**
3. Alignment label set: **Major and Minor only**
4. Design speed: **55mi/hr**
5. Minimum radius table: **AASHTO 2001 eMax 10%**
6. Save the file as *c05_ex01a.dwg*.

Chapter 6

Working with Profiles

Learning Objectives

After completing this chapter, you will be able to:

- *Understand the concept of profiles*
- *Create a surface profile*
- *Create a layout profile*
- *Create multiple, stacked, and quick profiles*
- *Create profile label styles and view styles*
- *Create band sets and band styles*
- *Use profile properties*

PROFILES

Alignments and profiles are the main components of a road design project. Profiles are graphical representations of the surface elevations along the horizontal alignment, as shown in Figure 6-1. They help you view the terrain along a specified alignment. A profile view object is used to store and display the information about the profile. It displays and maintains the profiles along with its associated information such as the length, start and end points, and station and elevation data of the alignment that it represents. Moreover, a profile view object creates and controls the actual grid in which the profile data will be displayed along with the bands of data that you want to display.

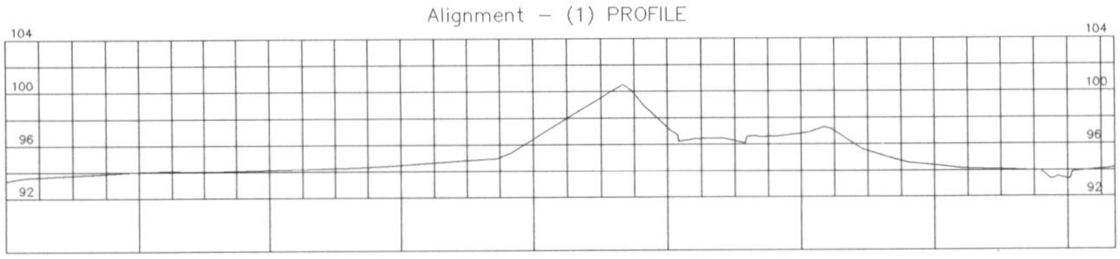

***Figure 6-1** A profile created from the alignment*

Creating a Surface Profile

Ribbon: Home > Create Design > Profile drop-down > Create Surface Profile
Command: CREATEPROFILEFROMSURFACE

You can create a surface profile and analyze the elevations along a horizontal path. To create a surface profile or an Existing Ground (EG) profile, choose the **Create Surface Profile** tool from the **Create Design** panel; the **Create Profile from Surface** dialog box will be displayed, as shown in Figure 6-2.

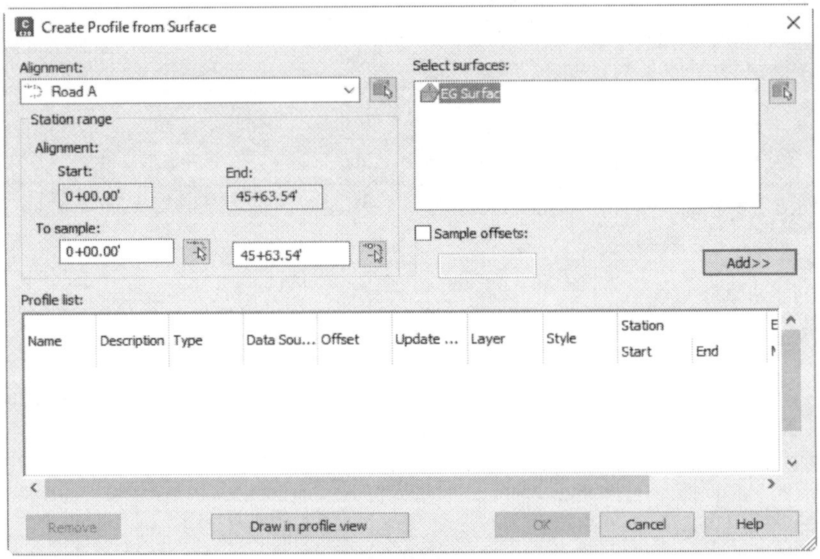

***Figure 6-2** The **Create Profile from Surface** dialog box*

Note

*The **Create Profile from Surface** dialog box will be invoked only if a surface object and an alignment object exists in the drawing.*

Creating a Profile View

To view a profile, it has to be drawn in the profile view. To create the profile view, choose the **Create Profile View** tool from **Home > Profile & Section Views > Profile View** drop-down; the **General** page of the **Create Profile View** wizard will be displayed, as shown in Figure 6-3. You will use this wizard to create the profile view.

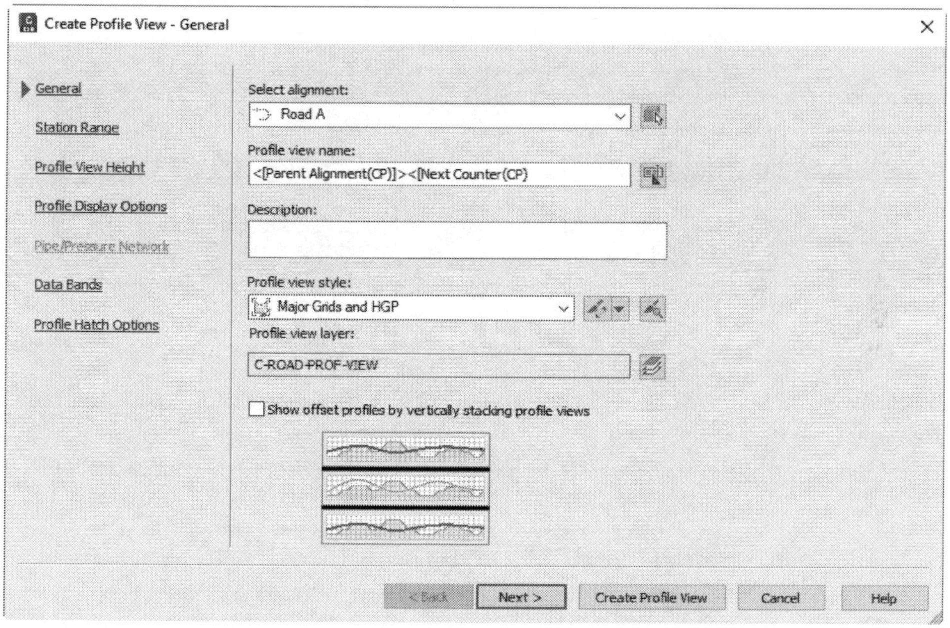

*Figure 6-3 The **General** page of the **Create Profile View** wizard*

Creating a Layout or Criteria-Based Profile

Ribbon: Home > Create Design > Profile drop-down > Profile Creation Tools
Command: CREATEPROFILELAYOUT

The layout profiles are used to represent the design elevations along an alignment and are created on the existing profile grid view. A layout profile represents the proposed elevation changes to be made for a finished ground. The workflow to create a layout (Design profile) is similar to creating a criteria-based alignment using the **Profile Layout Tools** toolbar. You need to select and assign a design criteria file for the profile from which you can obtain the minimum **K** values. Remember that if you assign a design criteria file to the parent alignment, the same design criteria file will be assigned to the profile view. However, you can change the design criteria file to create the profile as per your design requirements.

To create a layout profile, choose the **Profile Creation Tools** tool from the **Create Design** panel; you will be prompted to select a profile view. Select the profile view on which you want to create the design profile; the **Create Profile - Draw New** dialog box will be displayed, as shown in Figure 6-4.

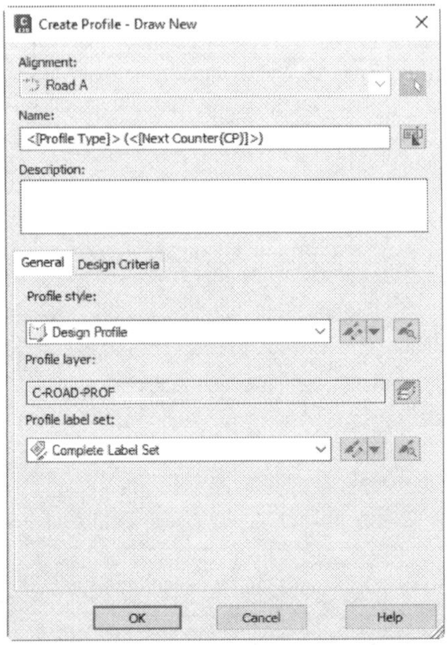

Figure 6-4 *The **Create Profile - Draw New** dialog box*

Creating a Superimposed Profile

Ribbon: Home > Create Design > Profile drop-down > Create Superimposed Profile
Command: SUPERIMPOSEPROFILE

In AutoCAD Civil 3D, you can compare two profiles by superimposing one profile over the other. A superimposed profile is the one that is created by superimposing the profile of an alignment over the profile views of other alignments. Superimposing a profile helps you study the variations in elevation of profiles. For example, you can superimpose a highway alignment profile over the pipe network alignment to view and compare the elevations.

To create a superimposed profile, choose the **Create Superimposed Profile** tool from the **Create Design** panel; you will be prompted to select the source profile from the drawing area. Select the profile which you want to superimpose as the source profile; you will be prompted to select the destination profile view where the superimposed profile is displayed. Select the profile view; the **Superimpose Profile Options** dialog box will be displayed, as shown in Figure 6-5.

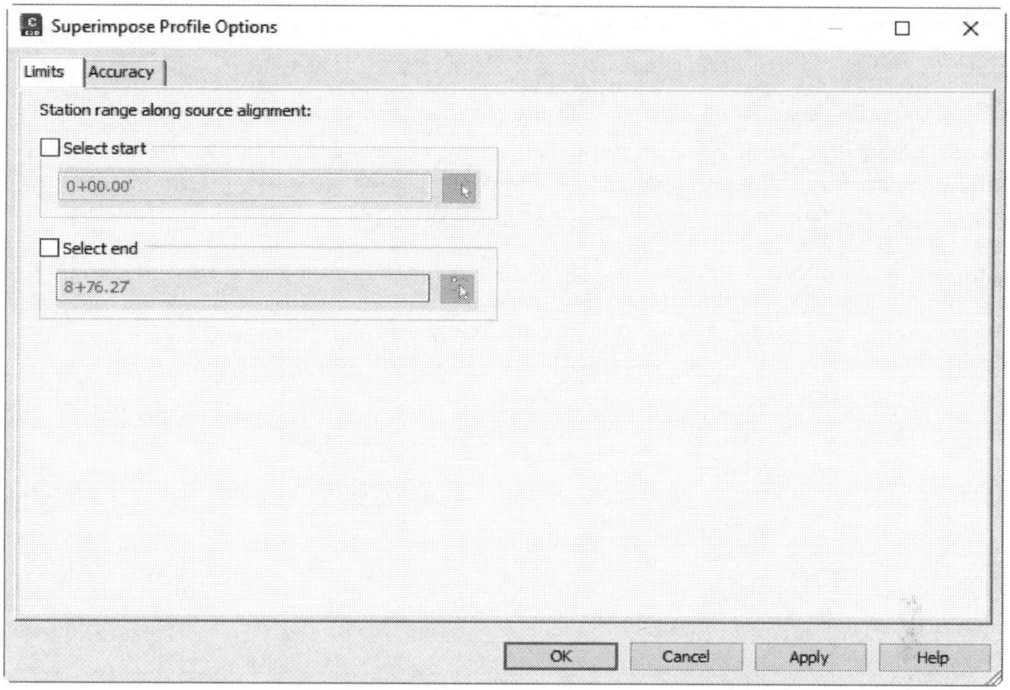

***Figure 6-5** The **Superimpose Profile Options** dialog box*

Creating Multiple Profile Views

Ribbon: Home > Profile & Section Views > Profile View drop-down > Create Multiple Profile Views
Command: CREATEMULTIPLEPROFILEVIEW

AutoCAD Civil 3D allows you to create multiple profile views by splitting the large profile view into smaller segments. To create multiple views, choose the **Create Multiple Profile Views** from the **Profile & Section Views** panel. On doing so, the **Create Multiple Profile Views** wizard along with the **General** page will be displayed, as shown in Figure 6-6.

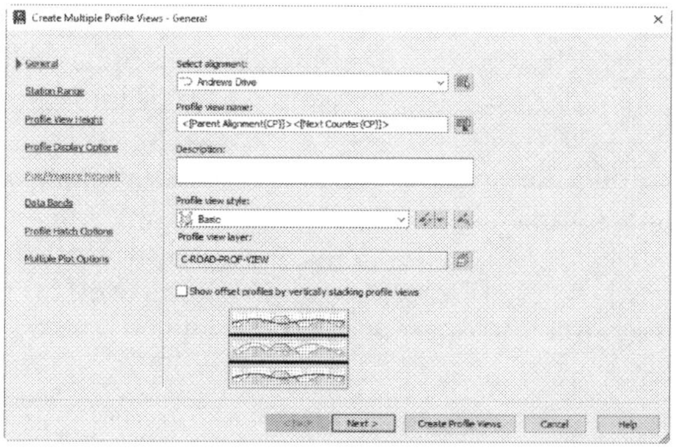

***Figure 6-6** The **General** page of the **Create Multiple Profile Views** wizard*

Creating a Quick Profile

Ribbon: Home > Create Design > Profile drop-down > Quick Profile
Command: CREATEQUICKPROFILE

You can create a profile without creating an alignment by using the **Quick Profile** tool. A quick profile can be created along an existing AutoCAD object such as line, polyline, feature line, lot line or using points that are specified by clicking on the screen.

To create a quick profile, choose the **Quick Profile** tool from the **Create Design** panel; you will be prompted to select the object along which you want to create the profile. Select the required object; the **Create Quick Profiles** dialog box will be displayed, as shown in Figure 6-7.

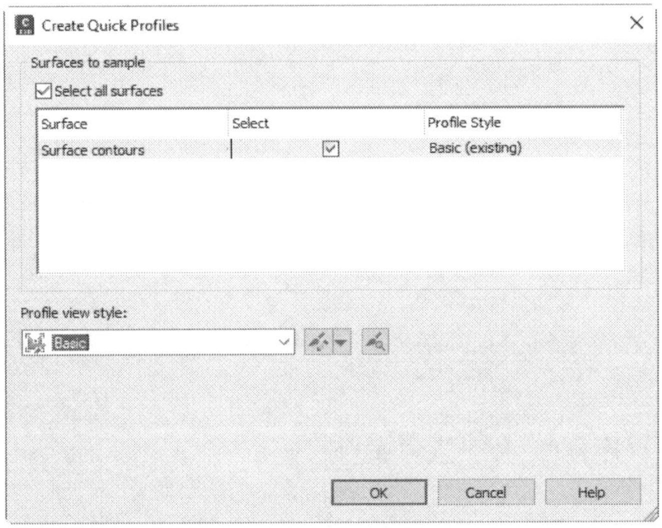

Figure 6-7 *The* ***Create Quick Profiles*** *dialog box*

Creating Dynamic Offset Profiles

You can create the dynamic offset profiles using the same command that you can use to create offset alignments. The geometry of the offset profile is created using a default cross slope which you can modify by editing the profile properties. You can create multiple offset alignments in a single operation including a different number on each side of the parent alignment. The offset distance can be different on each side of the parent alignment. It is used to show a borrow ditch, TBC profiles, or pretty much anything that is relative to your main profile.

To create an offset profile, choose the **Create Offset Alignments** tool from the **Alignments** drop-down in the **Create Design** panel; you will be prompted to select the alignment along which you want to create the offset profile. Select the required alignment; the **Create Offset Alignments** dialog box will be displayed. In this dialog box, select the **Create Profile for Offset Alignment** check box in the **Create Offset Profile** tab; the offset profile will be created, as shown in Figure 6-8.

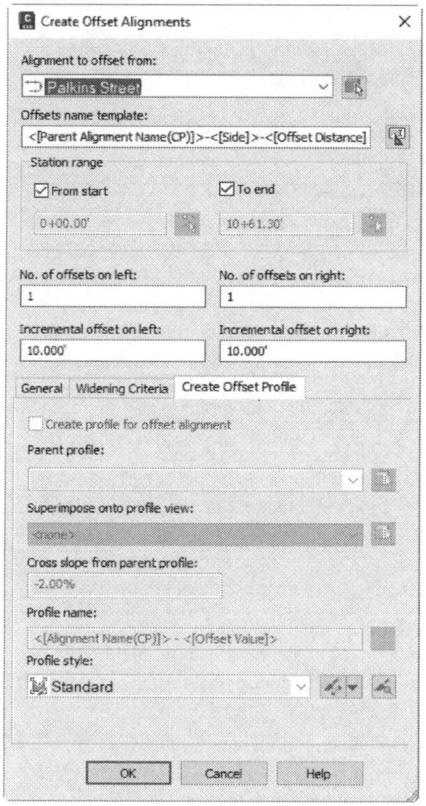

Figure 6-8 *The **Create Offset Profile** tab in the **Create Offset Alignment** dialog box*

DEFINING A PROFILE DESIGN CRITERIA

Ribbon: Profile (Contextual tab) > Modify Profile > Design Criteria Editor
Command: DESIGNCRITERIAEDITOR

The **Design Criteria Editor** tool is used to define or edit the design criteria for a design based profile. The **Design Criteria Editor** tool is available in the **Profile** (Contextual) tab. To create a new design criteria, choose the **Profile** tool from the **Design** panel of the **Modify** tab; the **Profile** tab will be displayed. Next, choose the **Design Criteria Editor** tool from the **Modify Profile** panel of the **Profile** tab; the **Design Criteria Editor - _Autodesk Civil 3D Imperial (2011) Roadway Design Standards.xml** dialog box will be displayed. Remember that the *_Autodesk Civil 3D Imperial (2011) Roadway Design Standards.xml* is the name of the default design criteria file that will be assigned to profiles. Before you create a new design criterion, choose the **Save As** button and save the default design criteria file with a different name at a different location. Now, expand the **Profiles > Minimum K Tables > AASHTO 2011 Standard** node to display three options for the **Stopping Sight**, **Passing Sight**, and **Headlight Sight Distances** under it. Select any of the options; the **Speeds** and **K** values will be displayed in the columns on the right side of the dialog box, as shown in Figure 6-9.

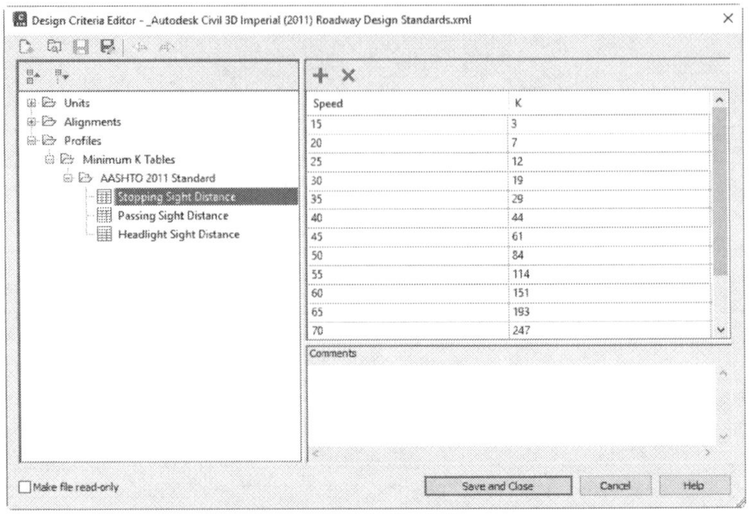

Figure 6-9 The ***Design Criteria Editor*** *dialog box showing the minimum K values at different speeds*

EDITING LAYOUT/DESIGN PROFILES

You can edit the design profile geometry and profile design criteria of a project at any stage. Civil 3D is compatible enough to quickly adjust to the changes made in the profile. You can edit the geometry of a profile either by grip editing or by using the **Profile Layout Parameters** dialog box, the **PANORAMA** window, and various other tools of the **Profile Layout Tools** toolbar. The profiles layout tools are described next.

Editing Profile Using the Profile Layout Tools Toolbar

The profile layout tools are used to modify the profile by editing the PVIs and profile entities individually. To do so, choose the **Profile** tool from the **Design** panel of the **Modify** tab; the **Profile** tab will be displayed in the Ribbon. In this tab, choose the **Geometry Editor** tool from the **Modify Profile** panel; you will be prompted to select the profile. Select the required profile from the drawing; the **Profile Layout Tools <layout name>** toolbar will be displayed.

To edit the PVIs in the layout profile, ensure that the **PVI based** tool is chosen in the **Profile Layout Tools** toolbar so that the PVI data of the selected profile is displayed in the **PANORAMA** window. To invoke the **PANORAMA** window, choose the **Profile Grid View** tool from the toolbar. The **PANORAMA** window contains the **Profile Entities** tab which lists all PVIs of the selected profile, refer to Figure 6-10.

Tip

You can also display the ***Profile Layout Tools <layout name>*** *toolbar from the drawing area. To do so, select the required layout profile from the profile view and right-click; a shortcut menu will be displayed. Choose* ***Edit Profile Geometry*** *from the shortcut menu; a toolbar will be displayed.*

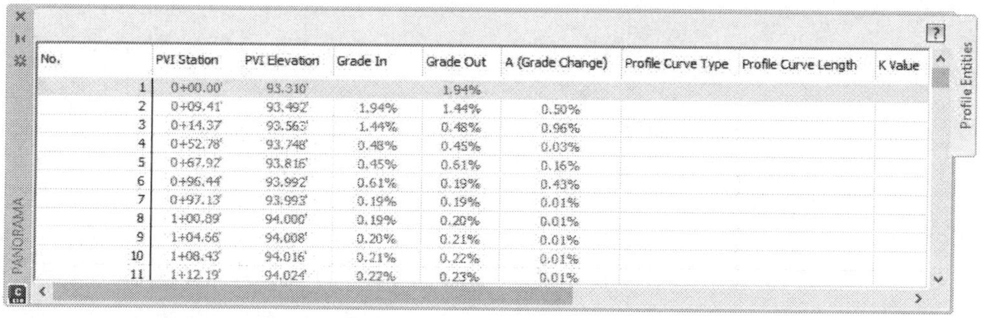

No.	PVI Station	PVI Elevation	Grade In	Grade Out	A (Grade Change)	Profile Curve Type	Profile Curve Length	K Value
1	0+00.00'	93.310'		1.94%				
2	0+09.41'	93.492'	1.94%	1.44%	0.50%			
3	0+14.37'	93.563'	1.44%	0.48%	0.96%			
4	0+52.78'	93.748'	0.48%	0.45%	0.03%			
5	0+67.92'	93.816'	0.45%	0.61%	0.16%			
6	0+96.44'	93.992'	0.61%	0.19%	0.43%			
7	0+97.13'	93.993'	0.19%	0.19%	0.01%			
8	1+00.89'	94.000'	0.19%	0.20%	0.01%			
9	1+04.66'	94.008'	0.20%	0.21%	0.01%			
10	1+08.43'	94.016'	0.21%	0.22%	0.01%			
11	1+12.19'	94.024'	0.22%	0.23%	0.01%			

Figure 6-10 *The* ***Profile Entities*** *tab of the* ***PANORAMA*** *window*

Editing Profiles Using Grips

One of the easiest methods of editing the geometry of a profile is by using the grips that are displayed on selecting a layout profile. You can edit both the profile PVIs and profile entities using the grips displayed on selecting the design profile. Before you do so, ensure that you have invoked the required **PVI Based** or **Entity Based** tool from the **Profile Layout Tools** toolbar. Now, select the required profile from the profile view to activate profile grips; the square, circular, and triangular shaped grips will be displayed.

PROFILE AND PROFILE VIEW STYLES

You can control the appearance of a profile and its components by creating a profile style. You must remember that the profile styles and profile view styles are two separate entities. The profile styles and profile view styles are created and managed in the **Settings** tab of the **TOOLSPACE** palette. The methods of creating styles for profile and profile views are discussed next.

Creating a Profile Style

To create a new profile style, choose the **Settings** tab of the **TOOLSPACE** palette. Expand **Profile > Profile Styles** subnode to view the default profile styles. Now, select and right-click on **Profile Styles** to display a shortcut menu. Choose **New** from the shortcut menu; the **Profile Style - New Profile Style** dialog box will be displayed, as shown in Figure 6-11.

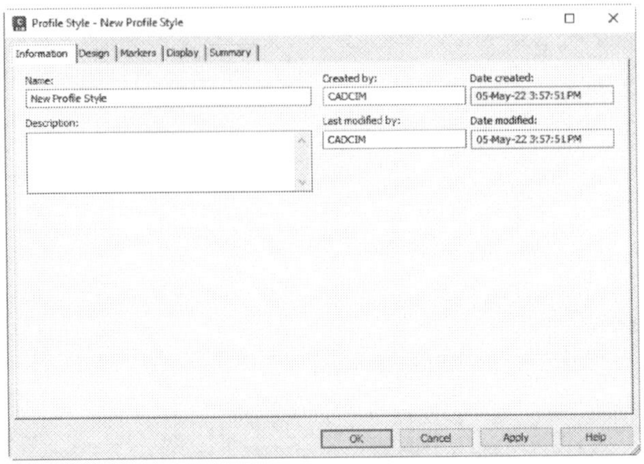

*Figure 6-11 The **Profile Style - New Profile Style** dialog box*

Creating a Profile View Style

The profile view style controls the display of the grid in which the profile is drawn. The profile view styles also control the grid and profile annotations. Like the profile style, the profile view styles are also created and managed in the **Settings** tab.

To create a profile view style, expand **Profile View > Profile View Styles** in the **Settings** tab; the existing profile view styles will be displayed. Next, right-click on the **Profile View Styles**; a shortcut menu will be displayed. Choose **New** from the shortcut menu; the **Profile View Style- New Profile View Style** dialog box will be displayed.

You can use this dialog box to create a customized profile view style by changing the default settings in different tabs of this dialog box.

Assigning the Profile View Style

You can assign a profile view style to the profile view by selecting the desired style from the **Profile view style** drop-down list available in the **General** page of the **Create Profile View** wizard, as shown in Figure 6-12.

Note

*You can access the **Create Profile View** wizard only if you create the profile by using the **Create Surface Profile** tool from the **Create Design** panel.*

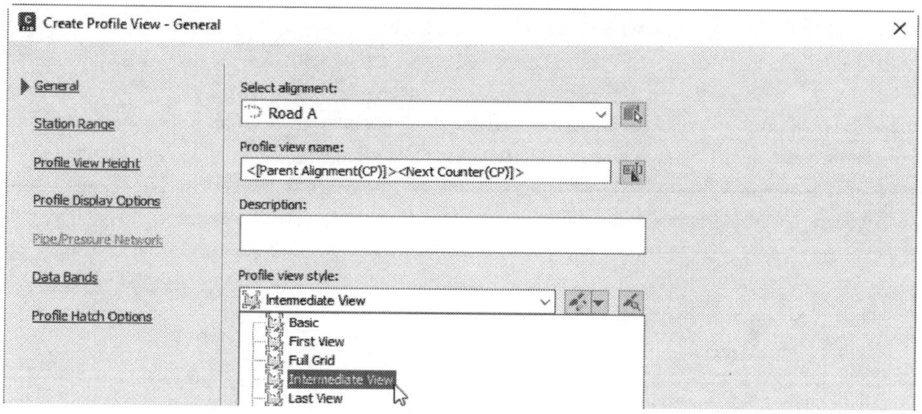

Figure 6-12 Selecting the intermediate view from the ***Profile view style*** *drop-down list in the* ***General*** *page of the* ***Create Profile View*** *wizard*

PROFILE LABELS

You can label a profile by using the in-built profile label styles. The layout profiles can be labeled automatically using the **Create Profile - Draw New** dialog box as explained earlier in this chapter. However, you need to use the **Edit Labels** command to label the other profiles. To label the layout profiles, select the **Complete Label Set** option from the **Profile label set** drop-down list of the **Create Profile - Draw New** dialog box, as shown in Figure 6-13.

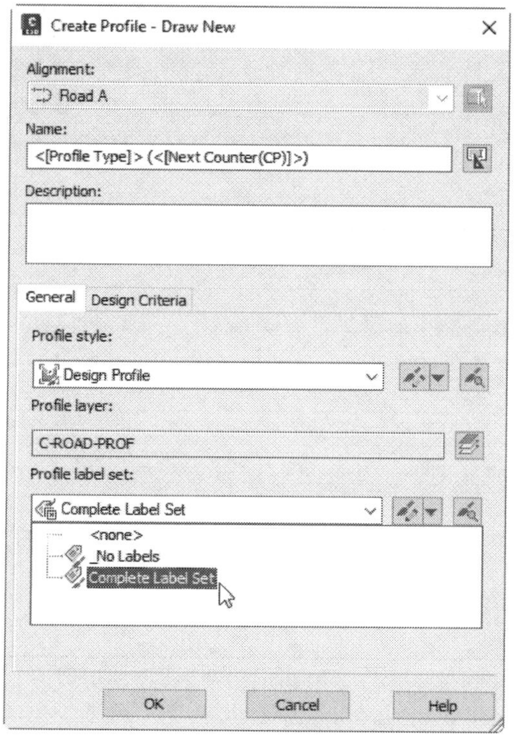

Figure 6-13 Selecting the ***Complete Label Set*** *from the* ***Profile label set*** *drop-down list*

Creating a New Profile Label Style

You can label the major and minor stations, sag and crest curves, horizontal geometry points, grade breaks, and lines in a profile view. You can also create your own label style to label all these components and then add them to label set. To create a new label for a curve, expand **Profile > Label Styles** node from the **Settings** tab. Next, select and right-click on **Curve** in the **Label Style** subnode to display the shortcut menu. Choose **New** from the shortcut menu; the **Label Style Composer - New Profile Curve Label Style** dialog box will be displayed, as shown in Figure 6-14.

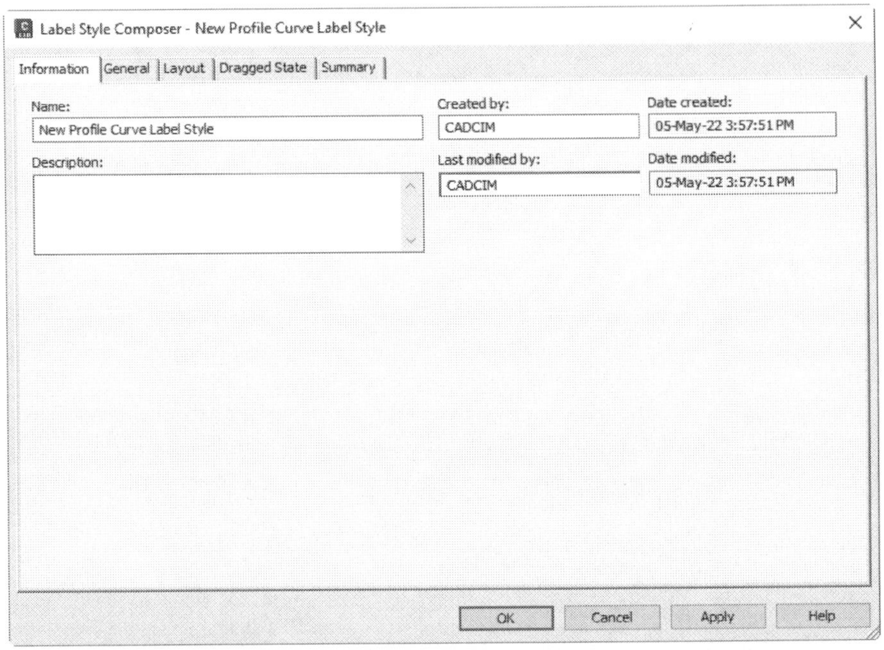

*Figure 6-14 The **Label Style Composer - New Profile Curve Label Style** dialog box*

Creating a Label Set

To create a label set, expand the **Profile > Label Styles** subnode in the **Settings** tab. Right-click on **Label Sets** in the **Label Styles** subnode of the **Profile** node; a shortcut menu will be displayed. Choose **New** from the shortcut menu; the **Profile Label Set - New Profile Label Set** dialog box will be displayed.

Assigning Labels to a Profile

Ribbon:	Annotate > Labels & Tables > Add Labels drop-down > Add/Edit Profile Labels
Command:	EDITPROFILELABELS

You can assign labels to a profile view as a label set. To do so, choose **Add/Edit Profile Labels** from the **Labels & Tables** panel; you will be prompted to select a profile from the drawing. Select the required profile line in the profile view grid; the **Profile Labels - <profile name>** dialog box will be displayed.

Creating a New Profile View Label Style

It is important to understand that the profile and profile views are two different things. Profiles are represented by a red line in the profile view. The profile view is the grid on which the profile is drawn. These styles can be accessed from the **Profile** and **Profile View** nodes in the **Settings** tab.

BAND SETS AND BAND STYLES

Civil 3D allows you to add profile data such as profile, superelevation, cut and fill depths, vertical geometry, and so on in a consolidated form of band in the profile view. Band set is a collection of such profile data bands and band styles. Figure 6-15 shows a profile view with a band set at the bottom of the profile view.

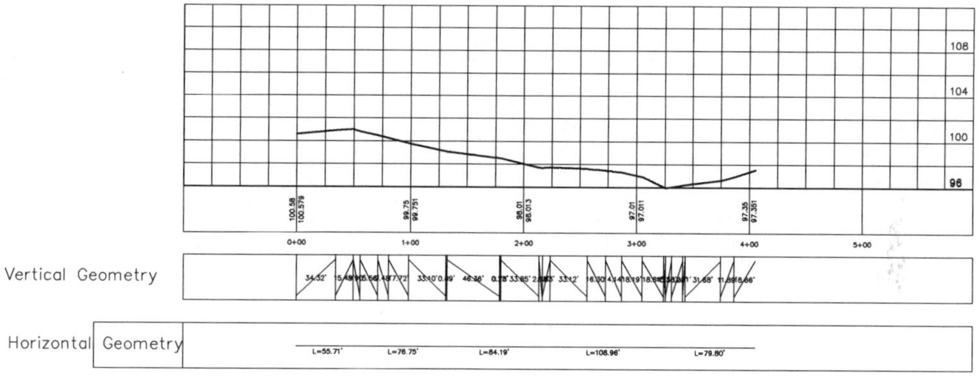

***Figure 6-15** Profile view with the band set at the bottom*

Creating a Band Style (Framing a Band)

You can create a new band style for each individual band type component and add it to the band set. For example, to create a new style for the display of the vertical geometry data, expand **Profile View > Band Styles > Vertical Geometry** in the **Settings** tab to view the in-built style for the **Vertical Geometry**. To create a customized style, right-click on the **Vertical Geometry** node; a shortcut menu will be displayed. Choose **New** from the shortcut menu; the **Vertical Geometry Band Style - New Vertical Geometry Band Style** dialog box will be displayed.

PROFILE PROPERTIES

Ribbon: Profile > Modify Profile > Profile Properties Drop Down> Profile Properties
Command: EDITPROFILEPROPERTIES

You can view the profile properties and the design criteria of the profile using the **Profile Properties** dialog box. To display the **Profile Properties** dialog box, choose the **Profile Properties** tool from the **Modify Profile** panel and select a profile from the drawing and press ENTER; the **Profile Properties - <profile name>** dialog box will be displayed, as shown in Figure 6-16.

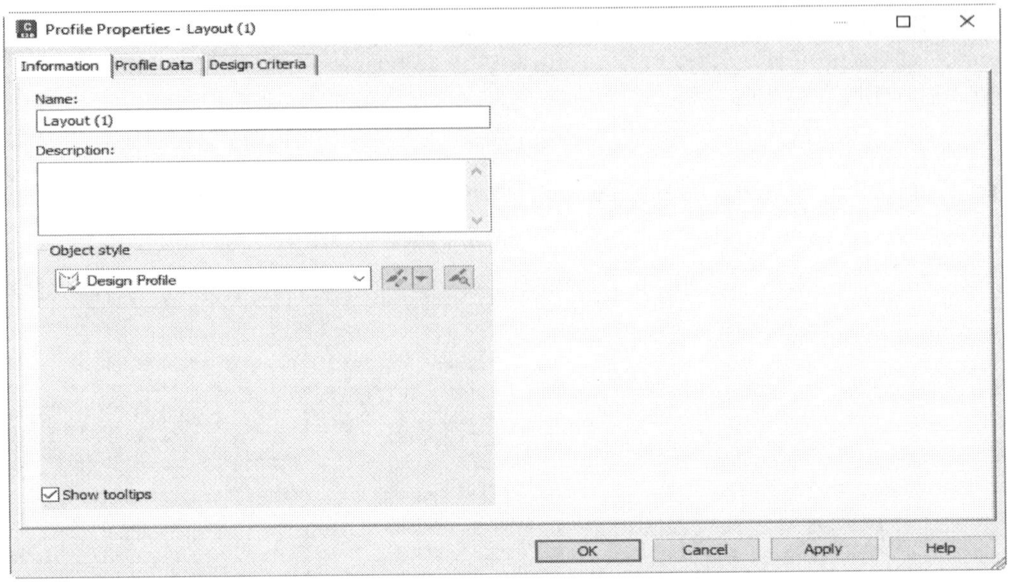

*Figure 6-16 The **Profile Properties - Layout (1)** dialog box*

PROFILE VIEW PROPERTIES

Ribbon:	Profile > Modify View > Profile View Properties Drop down > Profile View Properties
Command:	EDITGRAPHPROPERTIES

A profile view also has its own properties. To view or edit the profile properties, select the required profile view grid in the drawing and right-click to display the shortcut menu. Choose **Profile View Properties** from the shortcut menu; the **Profile View Properties - <profile view name>** dialog box will be displayed, as shown in Figure 6-17.

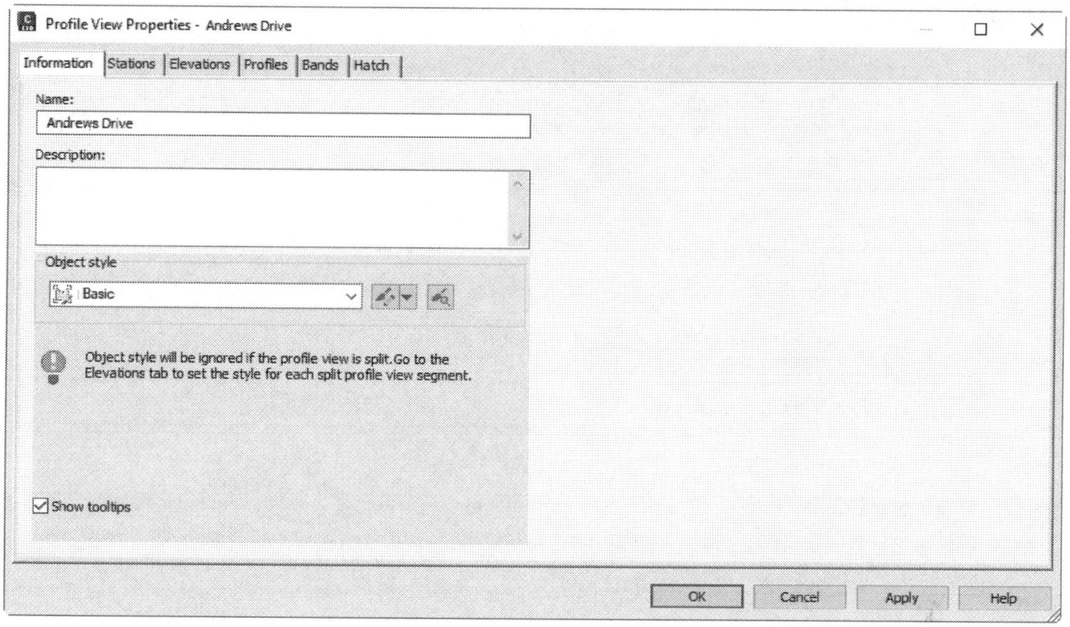

Figure 6-17 *The* ***Profile View Properties - Andrews Drive*** *dialog box*

TUTORIAL

Before starting the tutorial, you need to download and save the tutorial files on your computer. To do so, follow the steps given below:

1. Log on to *www.cadsofttech.com* and browse to *Books > Civil/GIS Books > AutoCAD Civil 3D > AutoCAD Civil 3D 2024 for Novices*. Next, select *c06_c3d_2024_tut.zip* file from the **Tutorial Files** drop-down list. Choose the **Download** button to download the data file.

2. Now, save and extract the downloaded folder to the following location:

   C:\c3d_2024

Note
While opening the tutorial file, the ***PANORAMA*** *window may appear with an error message. Close this window to proceed further.*

Tutorial 1 — Existing Ground Profile

In this tutorial, you will create a ground profile by sampling a surface across the given alignment. You will also create an offset profile and a staggered/split profile view. The split profile view is shown in Figure 6-18.

(Expected time: 30 min)

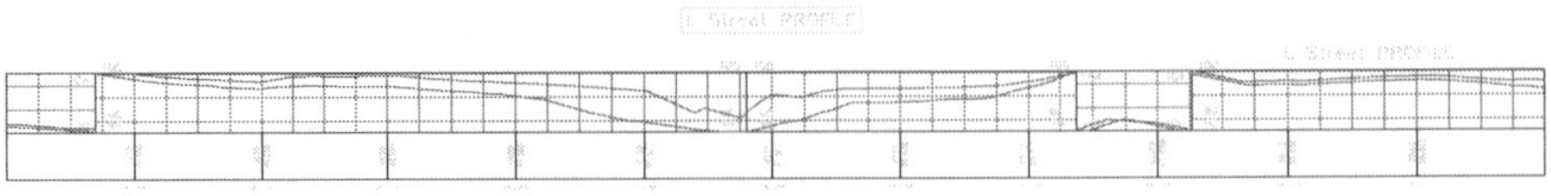

***Figure 6-18** The split profile view*

The following steps are required to complete this tutorial:

a. Open the file.
b. Create a profile using the **L Street** alignment.
c. Create the profile view.
d. Create the split/staggered profile view.
e. Save the file.

Opening the File

1. Open the *c06_c3d_2024_tut.zip* file that you downloaded from *https://www.cadsofttech.com* and saved at *C:\c3d_2024*.

2. Choose **Open** from the **Application Menu**; the **Select File** dialog box is displayed.

3. In the dialog box, browse to the location *C:\c3d_2024\c06_c3d_2024_tut,* where you have saved the file.

4. Select the file *c06_c3d_2024_tut01.dwg* and then choose the **Open** button to open the file. The file consists of a surface (**Surface1**) and two alignments, **L Street** and **L Street (1)**, as shown in Figure 6-19.

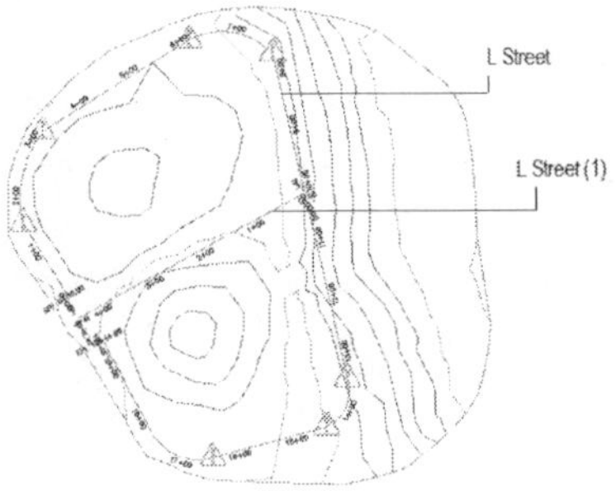

***Figure 6-19** The surface **(Surface1)** and the alignments displayed*

Creating a Profile

1. Choose the **Create Surface Profile** tool from **Home > Create Design > Profile** drop-down; the **Create Profile from Surface** dialog box is displayed.

2. In this dialog box, select the **L Street** option from the **Alignment** drop-down list if it is not selected by default.

3. Select **Surface1** from the **Select surfaces** list box and then choose the **Add** button; the selected profile is added to the **Profile list** list box.

4. Select the **Sample offsets** check box; the edit box below the **Sample offsets** check box is activated.

5. Enter **20** in the **Sample offsets** edit box and again select **Surface1** from the **Select surfaces** list box.

6. Choose the **Add** button; the offset profile is added to the **Profile list** list box.

Creating a Profile View

1. Choose the **Draw in profile view** button at the bottom of the **Create Profile from Surface** dialog box; the **General** page of the **Create Profile View** wizard is displayed.

2. Select **L Street** from the **Select alignmen**t drop-down list if it is not selected by default.

3. In the **Profile view name** edit box, enter **Street**.

4. Accept the default profile view style and choose the **Profile View Height** option from the left pane of the wizard; the **Profile View Height** page of the **Create Profile View** wizard is displayed.

5. In this page, select the **User specified** radio button from the **Profile view height** area of the **Profile View Height** page if not selected by default. Next, specify **97.00’** and **107.00’** in the **Minimum** and **Maximum** edit boxes, respectively, as shown in Figure 6-20.

Figure 6-20 *Specifying values for the Profile view height*

6. Next, choose the **Create Profile View** button from this page; you are prompted to select the origin of the profile view.

7. Click in the drawing area to place the profile view in the selected origin; the profile view and the offset are created. Figure 6-21 shows the **L Street** profile view created for the 20’ offset profile and the original profile.

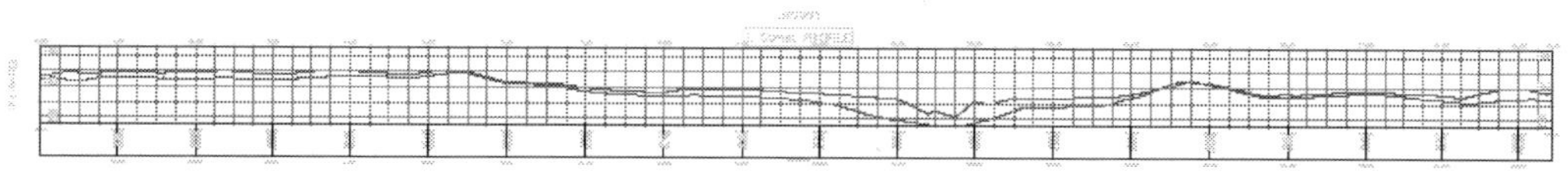

Figure 6-21 *The **L Street** profile view for the 20’ offset and the original profile*

The two profiles, the offset profile and the original profile are added to the **Profiles** node of the parent **L Street** alignment collection as **Surface1 - 20.000** (at 20' offset) and **Surface1 - Surface (17)** in the **Prospector** tab, as shown in Figure 6-22. To view the profile, expand **Alignments > L Street > Profiles** in the **Prospector** tab of the **TOOLSPACE** palette. The profile view is added to the **Profile Views** node as **Street**, as shown in Figure 6-23.

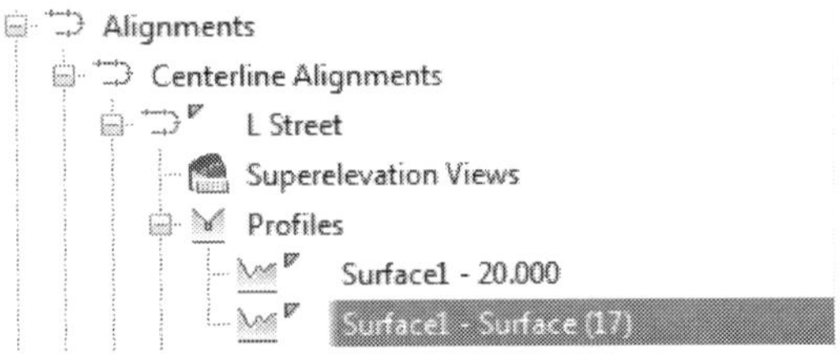

***Figure 6-22** Added profiles in the **Profiles** node of the **L Street** alignment*

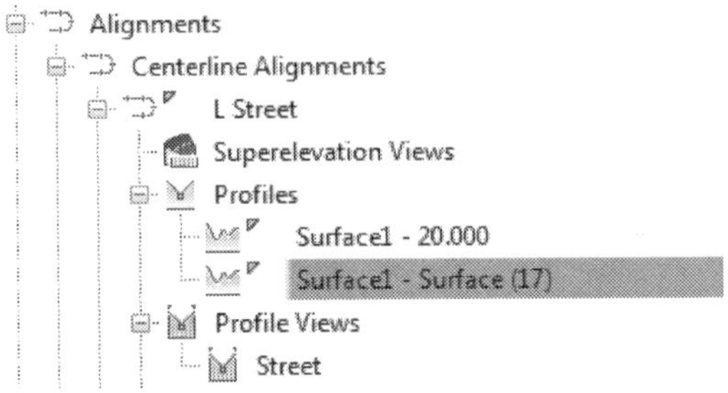

***Figure 6-23** The **Street** profile view in the **Profile Views** node of the **L Street** alignment*

Creating a Split/Staggered Profile View

1. Choose the **Create Profile View** tool from **Home > Profile & Section Views > Profile View** drop-down; the **General** page of the **Create Profile View** wizard is displayed. Ensure that **L Street** is selected in the **Select alignment** drop-down list.

2. Enter **View L** in the **Profile view name** edit box.

3. Select the **Full Grid** option from the **Profile view style** drop-down list if it is not selected by default.

4. Choose the **Next** button; the **Station Range** page is displayed.

5. In the **Station range** area, select the **User specified range** radio button if not selected by default.

6. Next, choose the button on the right of the first edit box in the **Station range** area to pick the first station from the drawing. 

7. Zoom in on the drawing and click at **6+00.00'** station to pick the first station.

> **Tip**
> *You can also edit the station range using the **Station Range** edit box. Click in the edit box and type in the required station value.*

8. Choose the button on the right of the second edit box and click at **18+00.00'** station tick mark to pick the end station to sample. 

9. Choose the **Next** button; the **Profile View Height** page is displayed.

10. In the **Profile View Height** area of this page; select the **User specified** radio button if not selected by default. Ensure that the **Split profile view** check box is selected.

11. Set the values in the **Minimum** and **Maximum** edit boxes on the right of the **User specified** radio button in the **Profile view height** area, as shown in Figure 6-24.

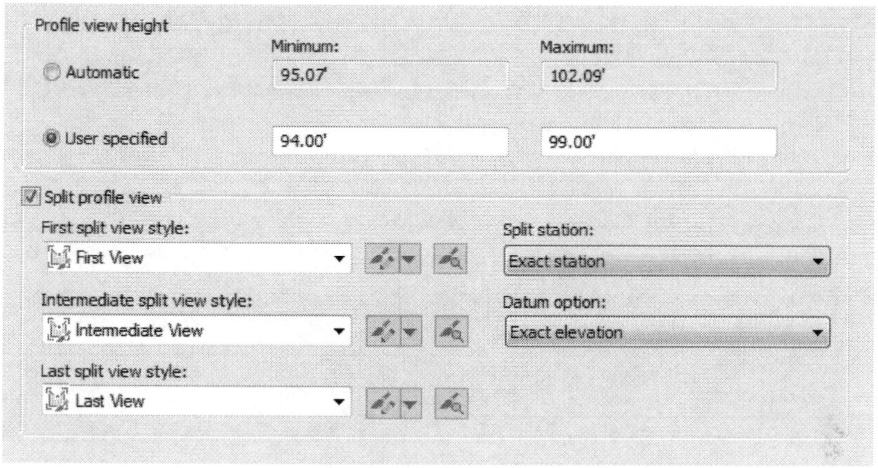

Figure 6-24 *Specifying the profile view height settings for split profiles*

12. Retain the default settings for the **Split profile view** area and choose the **Next** button in the **Create Profile View** wizard; the **Profile Display Options** page of the wizard is displayed. Make sure that the check box under the **Draw** column and the radio buttons in the **Clip Grid** and **Split At** columns corresponding to the **Surface1- Surface(17)** option in the **Name** column are selected.

13. Next, choose the **Create Profile View** button in the **Profile Display Options** page of the **Create Profile View** wizard; you are prompted to select the origin for the profile view.

14. Click in the drawing area to specify the origin of the profile view; the profile view is displayed and the profile is split in multiple views, as shown in Figure 6-25.

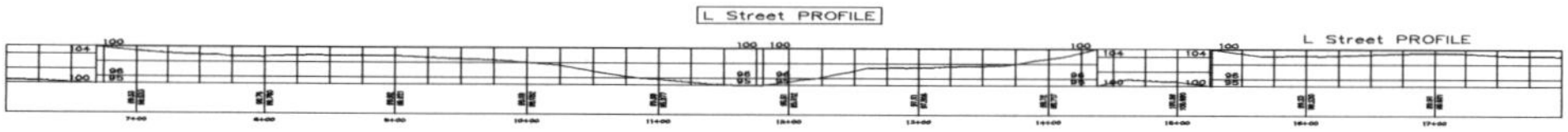

Figure 6-25 *The split profile view for the* ***L Street*** *alignment*

Saving the File

1. Choose **Save As** from the **Application Menu**; the **Save Drawing As** dialog box is displayed.

2. In this dialog box, browse to the following location:

C:\c3d_2024\c_06_c3d_2024_tut

3. In the **File name** edit box, enter **c06_tut01a**.

4. Next, choose the **Save** button; the file is saved with the name *c06_tut01a.dwg* at the specified location.

EXERCISES

1. Log on to *www.cadsofttech.com* and browse to *Books > Civil/GIS Books > AutoCAD Civil 3D > AutoCAD Civil 3D 2024 for Novices*. Next, select *c06_c3d_2024_exer.zip* file from the **Exercise Files** drop-down list. Choose the **Download** button to download the data file.

2. Now, save and extract the downloaded folder to the following location:

   C:\c3d_2024

Note

While opening the exercise files, the ***PANORAMA*** *window may appear with an error message. Close this window to proceed further.*

Exercise 1

Open the *c06_c3d_2024_ex01.dwg* file and create a split profile using the parameters given next.
(Expected time: 30 min)

1. Profile view name: **ST. Road**
2. Profile view height: Minimum: **94.483'** Maximum: **101.483'**
3. Split profile view: **Default settings**
4. Save the drawing as *c06_ex01a*.

Exercise 2

Open the *c06_c3d_2024_ex02.dwg* file and create a layout profile using the following tools and parameters:
(Expected time: 30 min)

1. Profile name: **Road A**
2. Profile style: **Default**
3. Profile label set: **Default**

Profile Layout Tools:

Curve: **Fixed Vertical Curve (Two points, parameter)**
Curve type: **Crest**
Curve radius: **35**

Chapter 7

Working with Assemblies and Subassemblies

Learning Objectives

After completing this chapter, you will be able to:

- *Understand and create assemblies*
- *Create different assembly styles*
- *Use codes*
- *Understand subassemblies*
- *Create customized subassemblies*

INTRODUCTION TO ASSEMBLIES AND SUBASSEMBLIES

After creating alignments and profiles, you will learn to create assemblies that are required to create a road model. A road model or corridor is formed by the combination of alignments, surface, and assemblies (discussed in next chapter). After designing the horizontal and vertical alignments of a roadway, you will now learn to define the cross-section of the roadway using assemblies.

An assembly is a collection of smaller units called subassemblies. These subassemblies are parametric in nature and are the building blocks of any road cross-section. A typical road section assembly consists of subassemblies such as lanes, sidewalks, and curbs. Figure 7-1 shows a road assembly cross-section.

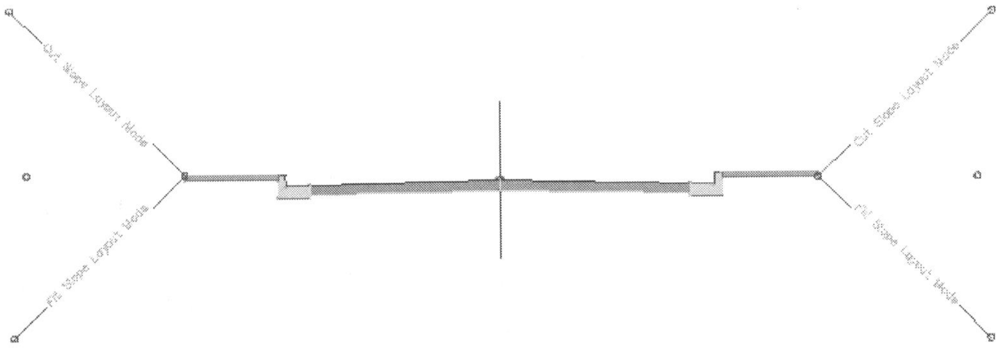

Figure 7-1 *A typical road assembly cross-section*

The subassemblies are attached to the left and right of the assembly and have some parameters associated with them. These parameters can be modified as per the requirements of the project. In short, assemblies together with subassemblies form the basic components of a road or other alignment based designs.

ASSEMBLIES

An assembly is a Civil 3D object that consists of subassemblies. The subassemblies are used to form the basic structure of a corridor. An assembly has a baseline with an attachment point indicated by a marker. A subassembly is attached to the baseline using this attachment point. Subassembly objects also have attachment points, that can be used to attach one subassembly to another.

Components of an Assembly

The primary components of an assembly are baseline and subassemblies. An assembly can also have a baseline point, insertion point, and offset line.

Creating Assemblies

Ribbon:	Home > Create Design > Assembly drop-down > Create Assembly
Command:	CREATEASSEMBLY

To create an assembly, choose the **Create Assembly** tool from the **Create Design** panel; the **Create Assembly** dialog box will be displayed, as shown in Figure 7-2.

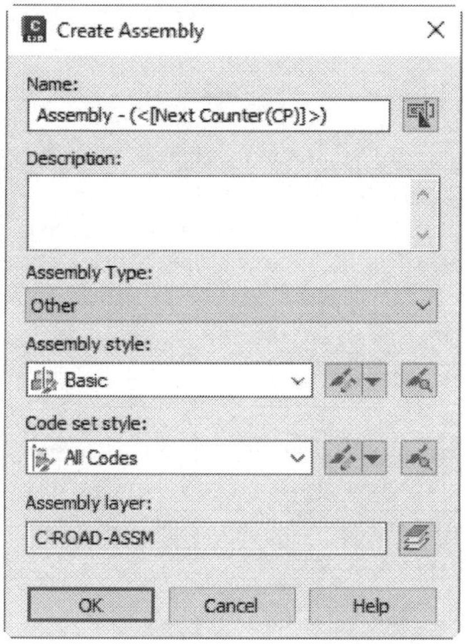

***Figure 7-2** The **Create Assembly** dialog box*

Assembly Properties

To access the assembly properties, choose the **Prospector** tab in the **TOOLSPACE**. Expand the **Assemblies** node. Right-click on the existing assembly in the node and choose the **Properties** option from the shortcut menu displayed; the **Assembly Properties - <Assembly Name>** dialog box will be displayed, refer to Figure 7-3.

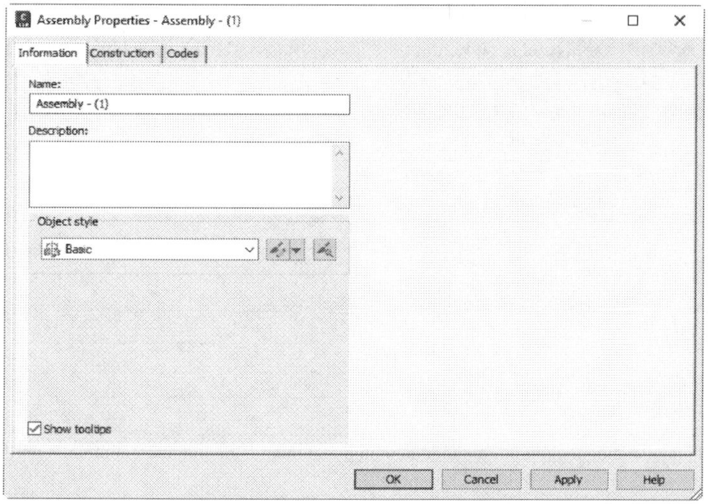

***Figure 7-3** The **Assembly Properties - Assembly - (1)** dialog box*

Assembly Styles

Like other Civil 3D objects, assemblies also have styles that control the display of the assemblies. The assembly styles are listed in the **Assembly** node of the **Settings** tab of the **TOOLSPACE**. You can create a new assembly style or edit the existing assembly style using the **Settings** tab.

USING CODES

Codes are used to display parameters of three main components of a subassembly namely, points, links, and shapes. These components are assigned with some values (annotation, style, labels) that are called codes. Codes identify the point, link, and shape in a subassembly and form the basic component of the three-dimensional corridor model. These codes also help create and arrange assemblies. There are three types of subassembly codes and they are discussed next.

Point Codes

Point codes define the vertices or the end points of a subassembly. These end points or vertices are called assembly points. For example, each crown point on a corridor will have the same point code and likewise, points on the edge of a pavement will have the same point code. These codes also help you create an offset or elevation label in road cross-sections. Figure 7-4 shows the points, links, and shapes in a subassembly.

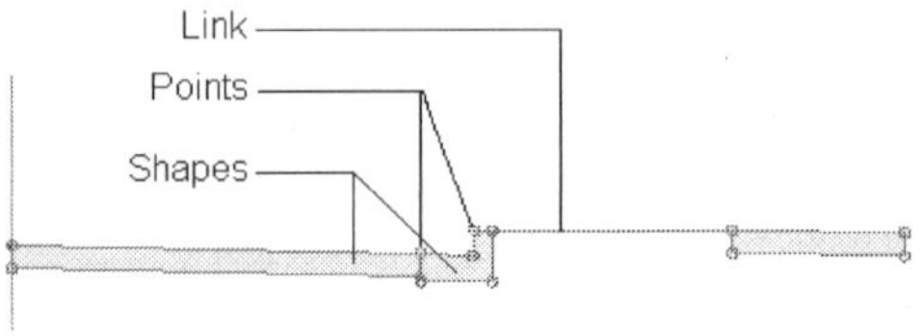

Figure 7-4 *Points, links, and shapes in a subassembly*

Link Codes

Links are basic lines that are used to join the points of a subassembly. These lines then connect with each other to form a planar structural surface of a corridor. Link codes are the codes assigned to the line segments joining the points. There are different types of link codes such as the Top link, Base link code, Datum link code, and Sub base link code. The Top link code connects all the points on the finished grade surface. Similarly, the Base link connects all the points located on the base layer. The link codes help you create grade or slope labels.

The link codes are used for tasks such as grading, visualization, and earthwork and volume calculations. You can use the links of a paved surface, finished grade surface, or sub-base surface and extract the required information from these links joining the point of a subassembly. In a subassembly, the points are represented by circles and the links are represented by straight lines joining the points.

Shape Codes

Shape of a subassembly is defined as a 2D closed area created by combining points and links. Shapes can represent a curb, sidewalk, pavement, and so on. The shape codes are used for quantity calculations based on the area of the shapes. Shape codes are used to identify the material and the hatch patterns.

Code Set Styles

Code set styles are similar to label set style or band set styles that have been explained in the earlier chapters. Code set style is a collection or a set of multiple code styles that are created and applied to objects once. You can create a customized style for links, points, or shape. You can add them in a code set and then add the styles together to a subassembly object. You can also create a label style for these components and add them to the code set. The code set styles are created and managed in the **Settings** tab of the **TOOLSPACE**. The code set styles are listed under the **Code Set Styles** subnode in the **Multipurpose Styles** node of the **General** node in the **Settings** tab, as shown in Figure 7-5. Before you create a code set style, create a style for different components of the subassembly such as point style, shape, and so on. You can then add these multiple styles in a code set to create a code set style.

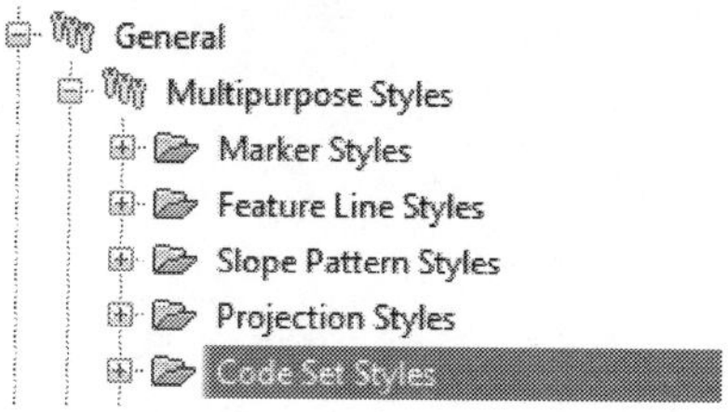

***Figure 7-5** The **Code Set Styles** subnode*

Creating Label Styles

You can also create label styles for point, link and shape components. The label styles control the display of labels and annotations used for these components. You can create label style for point, link, and shape. These labels represent the codes assigned to these components. They are created and managed in the **Settings** tab of the **TOOLSPACE**.

Creating Code Set Styles

To create a code set style, expand **General > Multipurpose Styles > Code Set Styles** subnode in the **Settings** tab of the **TOOLSPACE**. Next, right-click on the **Code Set Styles** node to display a shortcut menu and choose the **New** option from it; the **Code Set Style - New Code Set Style** dialog box will be displayed.

By default, the **Information** tab is chosen in this dialog box. In this tab, enter a name for the code set style in the **Name** edit box. Next, choose the **Codes** tab. This tab is used to view and change the default code styles as well as add new codes in the drawing. Code set styles can be associated with subassemblies, assemblies, corridors, and section views. Specify the code name, code style, and code label styles in the **Name**, **Style**, and **Label Style** columns, respectively, of the **Codes** tab, as shown in Figure 7-6.

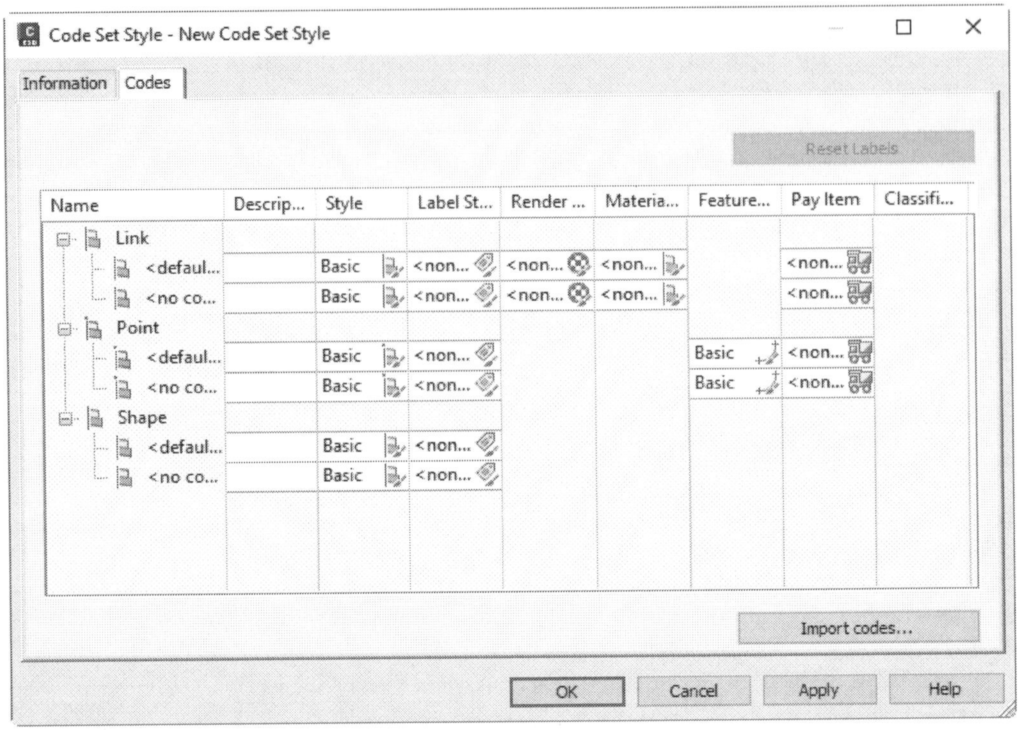

*Figure 7-6 The **Codes** tab of the **Code Set Style - New Code Set Style** dialog box*

WORKING WITH SUBASSEMBLIES

As explained earlier in this chapter, a subassembly is the main building block for corridor modeling and it represents the geometry of the components used for designing a corridor. AutoCAD Civil 3D has different in-built subassemblies for different components of a corridor such as lanes, ditches, and slopes. These ready-to-use subassemblies can adapt automatically to different superelevations, cut, or fill changes.

Besides the built in subassemblies, AutoCAD Civil 3D also allows you to create your own customized subassemblies using the polylines or .NET language. As discussed earlier, the subassemblies can be accessed from the **TOOL PALETTES**, as shown in Figure 7-16 in book. The **TOOL PALETTES** contains various tabs with different subassemblies. When you hover the cursor over a tab, a tooltip is displayed with the name of the tab and type of subassemblies.

Subassembly Properties

You can view the properties of the subassembly in the **Subassembly Properties** dialog box. To invoke this dialog box, select the subassembly in the drawing and right-click to display the shortcut menu. Choose **Subassembly Properties** from the shortcut menu; the **Subassembly Properties - <Subassembly Name>** dialog box will be displayed.

Creating Subassemblies using Polylines

You can create a customized subassembly using polylines. To do so, create the required object using the **Polyline** tool. Next, choose the **Create Subassembly from Polyline** tool from the **Create Design** panel of the **Home** tab; you will be prompted to select the entity. Select the polyline entity from the drawing area; the **Create Subassembly - From Polyline** dialog box will be displayed, as shown in Figure 7-7.

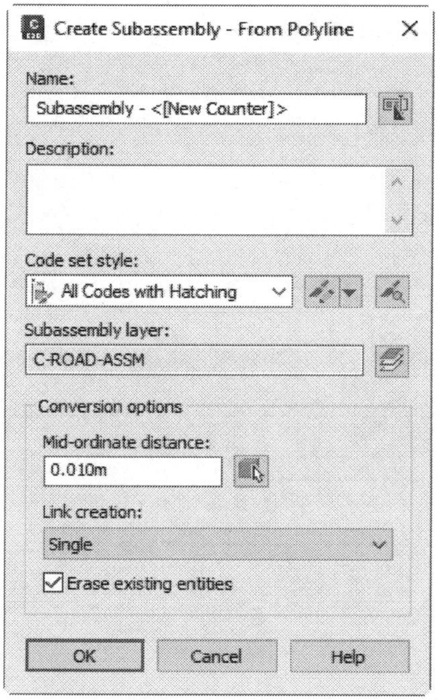

*Figure 7-7 The **Create Subassembly - From Polyline** dialog box*

Adding Codes to the Subassembly

After creating a subassembly from a polyline, you need to add codes to the subassembly. To add code, select the subassembly which is created from the polyline, and then right-click on it; a shortcut menu is displayed. Choose **Add Code** from the shortcut menu; you will be prompted to enter a code. Specify a code in the command line and press ENTER; you will be prompted to select a link, point, or shape. Select a point, link, or shape of the subassembly; the code will be assigned to the selected subassembly.

To view the assigned codes, select the assembly in the drawing and right-click to display the shortcut menu. Choose **Subassembly Properties** from the shortcut menu; the **Subassembly Properties - <assembly name>** dialog box will be displayed. Choose the **Codes** tab to view the assigned codes.

Subassembly Catalog Library

The subassembly catalog displays different subassembly tools that can be used for the corridor modeling. These subassemblies are created using the .NET programming language. To access the subassembly catalogs, select the assembly from the drawing; the **Sub-Assembly<name>** tab will be displayed. Choose the **Catalog** button from **Launch Pad** panel of this tab; the **<user name> Catalog Library - Autodesk Content Browser 2024** window will be displayed, as shown in Figure 7-8.

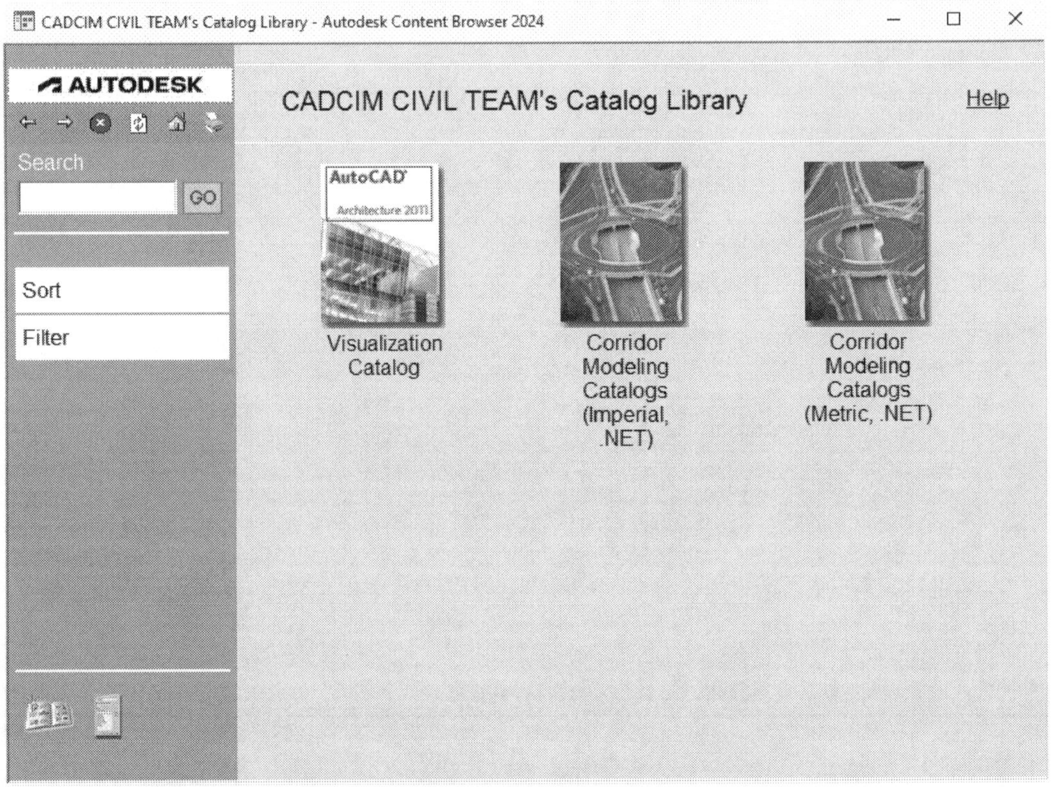

Figure 7-8 The window displaying the ***Catalog Library - Autodesk Content Browser 2024*** *window*

ADDING ASSEMBLIES TO TOOL PALETTES

Once you have created the assembly for your corridor model, you can save it in the **TOOL PALETTES**. To do so, enter **B** at the command prompt and press ENTER; the **Block Definition** dialog box will be displayed. You need to convert the entire assembly into one block. Enter a name for the assembly in the **Name** edit box. Choose the **Pick Point** button from the **Base point** area; the **Block Definition** dialog box will be closed and you will be prompted to specify the insertion base point. Select the insertion base point from the assembly by clicking on the assembly at the required point; the **Block Definition** dialog box will be displayed again and the coordinates of the selected insertion point will be displayed in the **Base point** area. Next, clear the **Specify On-screen** check box from the **Objects** area. Ensure that the **Open in block editor** check box is cleared so as to make the resulting block non-editable. Choose the **Select objects** button; the **Block Definition** dialog box will be closed and you will be prompted to select the objects. Select the complete assembly and the baseline with a crossing window. Press ENTER to return to the dialog box. Next, choose the **OK** button to close the **Block Definition** dialog box and convert the assembly into a block. Save the file at the required location. Next, enter **Insert** in the command line and press ENTER; the **Insert** dialog box will be displayed. Select the block from the **Name** drop-down list. Choose the **Browse** button and open the saved file from the **Select Drawing File** dialog box displayed. Select the **Explode** check box and choose the **OK** button; the **Select Drawing File** dialog box will be closed and you will be prompted to specify the insertion point. Click in the drawing at the suitable location. Specify the scale factor and press ENTER. The assembly block will be added to the drawing. You can zoom in or zoom out to view the assembly clearly. Save the file again.

TUTORIAL

Before starting the tutorial, you need to download and save the tutorial files on your computer. To do so, follow the steps given below:

1. Log on to *www.cadsofttech.com* and browse to *Books > Civil/GIS Books > AutoCAD Civil 3D > AutoCAD Civil 3D 2024 for Novices.* Next, select *c07_c3d_2024_tut.zip* file from the **Tutorial Files** drop-down list. Choose the **Download** button to download the data file.

2. Now, save and extract the downloaded folder to the following location:

   C:\c3d_2024

Note
While opening the tutorial file, the ***PANORAMA*** *window may appear with an error message. Close this window to proceed further.*

Tutorial 1 — Typical Road Assembly

In this tutorial, you will create a typical road assembly consisting of different subassemblies, as shown in Figure 7-9 and then rename the subassemblies in the created road assembly.

(Expected time: 30 min)

The following steps are required to complete this tutorial:

a. Open the downloaded file.
b. Create an assembly.
c. Add the required subassemblies after modifying their parameters.
d. Rename the subassemblies.
e. Save the file.

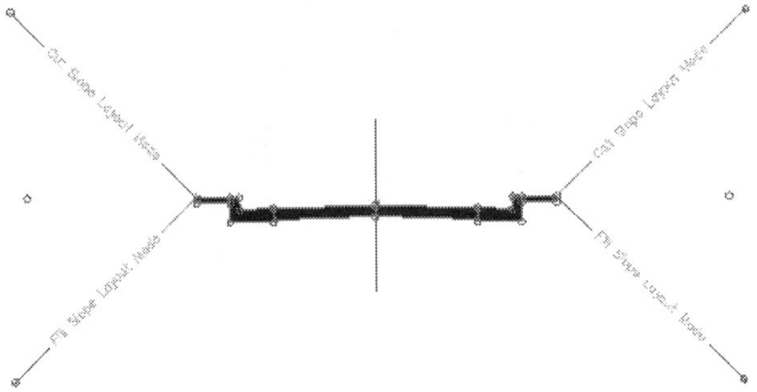

Figure 7-9 *A typical road assembly*

Opening the Downloaded File

1. Open the *c07_c3d_2024_tut.zip* file that you downloaded from *https://www.cadsofttech.com* and save at *C:\c3d_2024.*

2. Choose **Open** from the **Application Menu**; the **Select File** dialog box is displayed.

3. In this dialog box, browse to *C:\c3d_2024\c07_c3d_2024_tut* where you have saved the file.

4. Select the *c07_c3d_2024_tut01.dwg* file and then choose the **Open** button to open it.

Ignore any error message if displayed.

Creating an Assembly

1. Choose the **Create Assembly** tool from **Home > Create Design > Assembly** drop-down; the **Create Assembly** dialog box is displayed.

2. Enter **Typical Road** in the **Name** edit box.

3. Next, choose the **OK** button; the **Create Assembly** dialog box is closed and you are prompted to specify the assembly baseline location.

4. Click in the drawing area to specify the location for the assembly baseline; the assembly baseline is added, as shown in Figure 7-10.

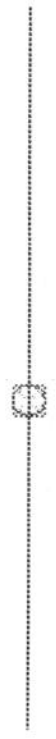

Figure 7-10 *The assembly baseline*

Adding Subassemblies

1. Choose the **Tool Palettes** button from the **Palettes** panel of the **Home** tab; the **TOOL PALETTES** is displayed.

2. Choose the **Basic** tab from the **TOOL PALETTES** and select the **BasicLaneTransition** subassembly from it; the **Properties** window is displayed.

3. Set the value of the **Default Width** parameter to **6.000'** in the **ADVANCED** rollout of the **Properties** window. Set the value of the **Side** property to **Right**.

Note

*You can press CTRL + 1 to display the **Properties** window if not displayed.*

4. Next, click on the marker of the assembly baseline; the **BasicLaneTransition** subassembly is added to the right of the baseline.

5. Set the **Side** parameter in the **Properties** window to **Left**, refer to Figure 7-11.

6. Ensure that the value of the **Default Width** property is **6.000'**.

7. Click on the assembly marker; the **BasicLaneTransition** subassembly is added to the left of the assembly baseline. Figure 7-12 shows the **BasicLaneTransition** subassembly added to both sides of the assembly baseline.

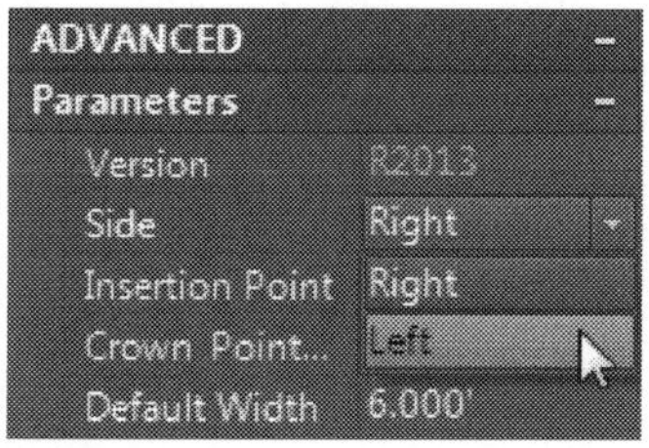

Figure 7-11 *Setting the **Side** parameter to **Left***

Note

*You must check the side in the **Properties** window and then click on the respective side of the assembly baseline to add the subassembly.*

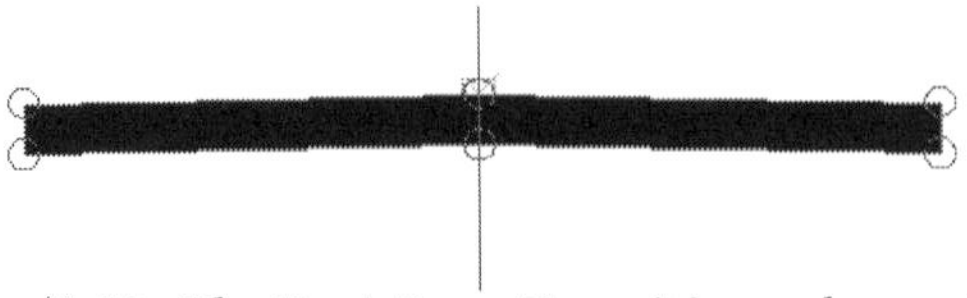

Figure 7-12 *The **BasicLaneTransition** subassemblies added to both sides of the assembly baseline*

8. In the **Basic** tab of the **TOOL PALETTES**, select the **BasicCurbAndGutter** subassembly; the **Properties** window for the subassembly is displayed. Ensure that the **Side** parameter is set to **Right** in the **ADVANCED** rollout of the **Properties** window. 

9. Click on the circular marker on the top right edge of the **BasicLaneTransition** subassembly, as shown in Figure 7-13; the **BasicCurbAndGutter** subassembly is added to the right of the **BasicLaneTransition** subassembly.

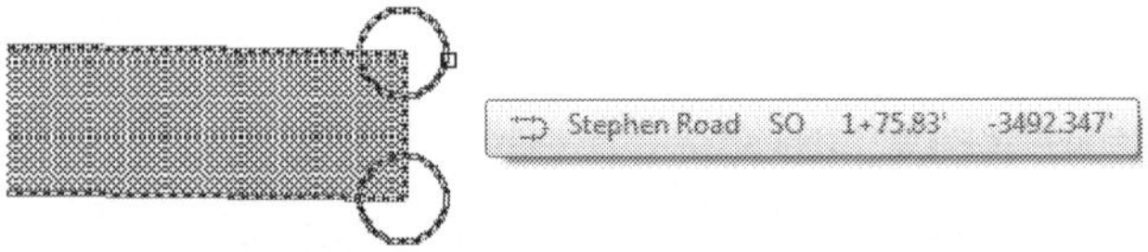

Figure 7-13 *Selecting the top circular mark of the **BasicLaneTransition** subassembly*

10. Next, set the **Side** parameter for the **BasicCurbAndGutter** subassembly to **Left** in the **Properties** window.

11. Click on the circular mark on the top left edge of the **BasicLaneTransition** subassembly. The **BasicCurbAndGutter** subassembly is added to the left of the assembly baseline.

    Now, you have added a curb and gutter to the pavement. Zoom out to see the entire cross section of the pavement, refer to Figure 7-14.

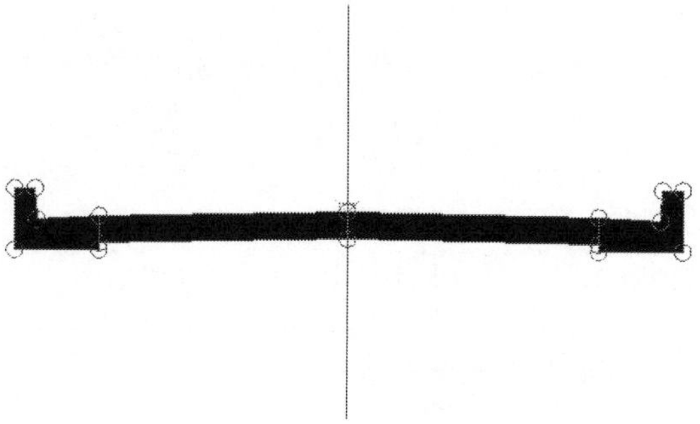

Figure 7-14 *The **BasicCurbAndGutter** subassembly added to the assembly*

12. Select the **BasicSidewalk** subassembly from the **Basic** tab in **TOOL PALETTES**. 

13. Set the **Width** parameter value to **2.000'** in the **Properties** window and press ENTER.

14. Set the **Side** parameter value to **Right** in the **Properties** window.

15. Next, click on the circular marker on the top right edge of the **BasicCurbAndGutter** subassembly, as shown in Figure 7-15; a 2' wide sidewalk is added to the right side of the assembly.

Figure 7-15 *Selecting the circular marker on the right side of the top edge of the **BasicCurbAndGutter** subassembly*

16. In the **TOOL PALETTES**, change the value of the **Side** parameter to **Left** and ensure that the **width** parameter is set to **2'**. Now, click on the circular marker on the top left edge of the **BasicCurbAndGutter** subassembly. A sidewalk of width 2' is added to left side of the assembly. Figure 7-16 displays **BasicSidewalk** subassembly added to both sides.

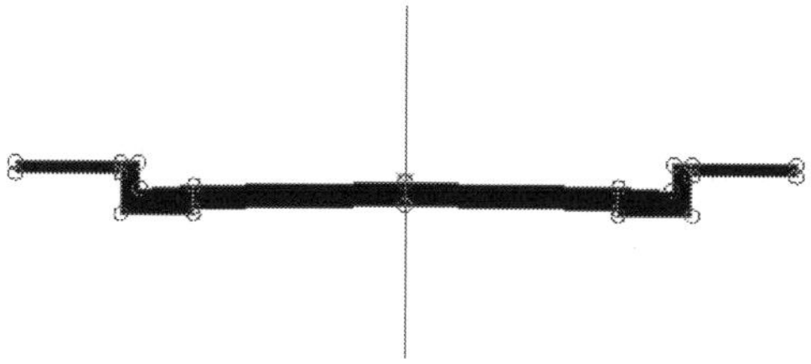

Figure 7-16 *The **BasicSidewalk** subassembly added to both sides of the assembly baseline*

17. Next, select the **BasicSideSlopeCutDitch** subassembly in the **Basic** tab of the **TOOL PALETTES**.

18. Set the **Cut Slope** parameter value to **4.00:1** in the **Properties** window and press ENTER.

19. Ensure that the **Side** parameter is set to **Right**.

20. Click on the marker on the top right edge of the **BasicSidewalk** subassembly; the **BasicSideSlopeCutDitch** subassembly is added to the right of the assembly.

21. Repeat steps 17 and 18, change the **Side** parameter to **Left** and add the **BasicSideSlopeCutDitch** subassembly to the left of the assembly. Now, you have finished creating a **Typical Road** assembly consisting of the **BasicLaneTransition**, **BasicCurbAndGutter**, **BasicSidewalk**, and **BasicSideSlopeCutDitch** subassemblies, as shown in Figure 7-17. Next, close the **Properties** window and the **TOOL PALETTES**, and press ESC to exit the command.

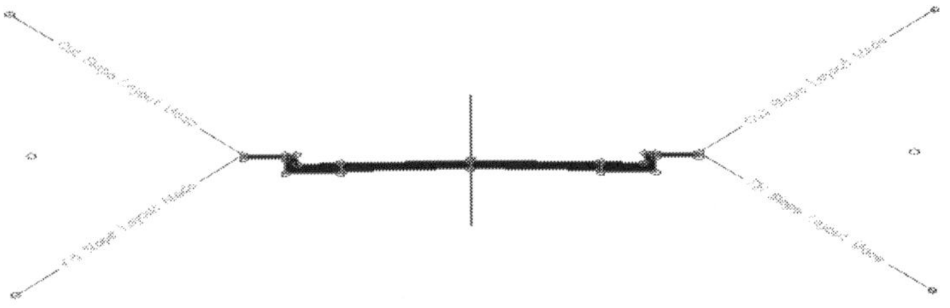

Figure 7-17 *A typical road assembly*

> **Tip**
> *To invoke the **Assembly Properties - <Assembly Name>** dialog box, select the required assembly baseline in the drawing and right-click; a shortcut menu is displayed. Choose **Assembly Properties** from the shortcut menu.*

Renaming the Subassemblies

1. Choose the **Prospector** tab of the **TOOLSPACE**.

2. Expand the **Assemblies** node.

3. Right-click on **Typical Road** in the **Assemblies** node; a shortcut menu is displayed.

4. Choose the **Properties** option from the shortcut menu; the **Assembly Properties - Typical Road** dialog box is displayed.

5. Choose the **Construction** tab; the subassemblies are listed under the **Item** area.

6. Right-click on the **Right** subnode in the **Item** area; a shortcut menu is displayed. Choose the **Rename** option from the shortcut menu; the selected text becomes editable.

7. Enter **Right Lane** in the displayed edit box and press ENTER.

8. Using the procedure followed in steps 6 and 7, rename the **Left** subnode as **Left Lane**.

9. Next, right-click on the **BasicLaneTransition** subassembly under the **Right Lane** and choose **Rename** from the shortcut menu displayed; the selected text becomes editable.

10. Enter **BasicLaneTransition - R** in the displayed edit box and press ENTER.

11. Next, right-click on the first **BasicLaneTransition** subassembly under the **Left Lane** subnode and choose **Rename** from the shortcut menu displayed; the subassembly name becomes editable.

12. Enter **BasicLaneTransition - L** and press ENTER. Similarly add suffixes **- R** and **- L** to the other subassembly names under the **Right Lane** and **Left Lane** subnodes, respectively, refer to Figure 7-18.

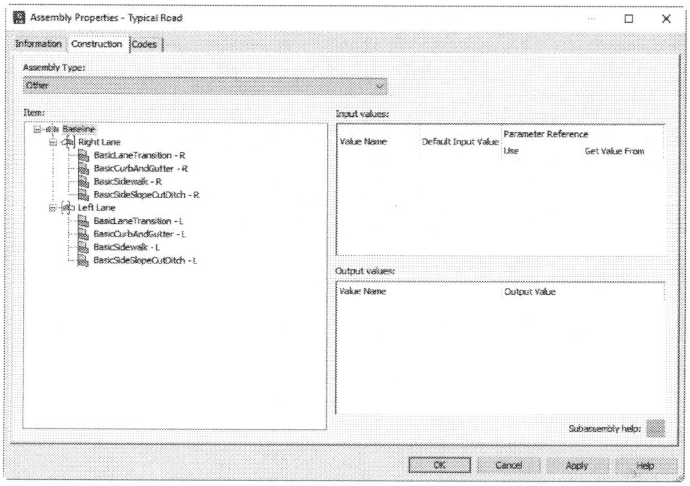

Figure 7-18 *The **Item** area of the **Construction** tab displaying the renamed subassemblies*

13. Next, choose the **OK** button in the **Assembly Properties - Typical Road**; the dialog box is closed.

14. Expand the **Assemblies** node and its subnodes in the **Prospector** tab; the subassemblies are renamed and displayed, as shown in Figure 7-19.

Note

*You can also rename the subassemblies by selecting each of them individually in the drawing and then renaming them in the **Subassembly Properties** dialog box. To display the **Subassembly Properties** dialog box, select the subassembly in the drawing, right-click on it and choose **Subassembly Properties** from the shortcut menu displayed.*

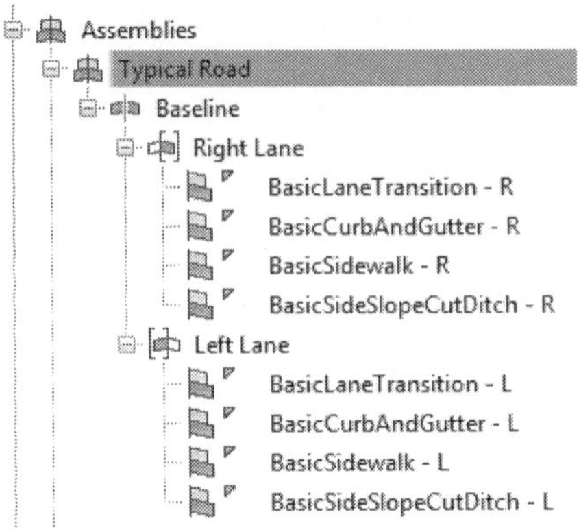

Figure 7-19 *The renamed subassemblies in the* ***Assemblies*** *node in the* ***Prospector*** *tab*

Saving the File

1. Choose **Save As** from the **Application Menu**; the **Save Drawing As** dialog box is displayed.

2. In this dialog box, browse to the following location:

   C:\c3d_2024\c07_c3d_2024_tut

3. In the **File name** edit box, enter **c07_tut01a**.

4. Choose the **Save** button; the file is saved with the name *c07_tut01a.dwg* at the specified location.

EXERCISES

1. Log on to *www.cadsofttech.com* and browse to *Books > Civil/GIS Books > AutoCAD Civil 3D > AutoCAD Civil 3D 2024 for Novices*. Next, select *c04_c3d_2024_exer.zip* file from the **Exercise Files** drop-down list. Choose the **Download** button to download the data file.

2. Now, save and extract the downloaded folder to the following location:
   C:\c3d_2024

Note
*While opening the exercise files, the **PANORAMA** window may appear with an error message. Close this window to proceed further.*

Exercise 1

Create a Road assembly, refer to Figure 7-20 using the following subassemblies and parameters: **(Expected time: 30 min)**

Road Assembly Name: **Perkins Road**
Subassembly 1: **BasicLaneTransition**
Default Width: **10.000'**

Subassembly 2: **BasicCurbAndGutter**
Gutter Width: **1.500'**

Subassembly 3: **UrbanSidewalk**
Sidewalk Width: **3.000'**
Subassembly 4: **SideDitch**
Inside Extension Height: **0.000'**
Inside Backslope Width: **0.000'**

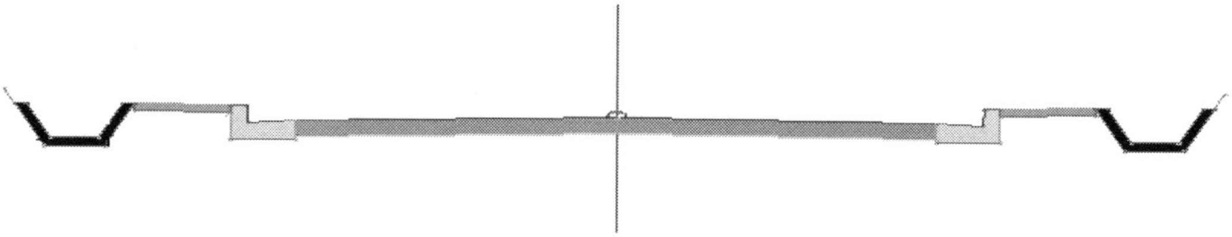

Figure 7-20 *The **Perkins Road** assembly*

Exercise 2

Create a pipe trench assembly, as shown in Figure 7-21 using the following subassembly and parameters: **(Expected time: 30 min)**

Assembly Name: **PipeTrench**
Subassembly: **TrenchPipe1**

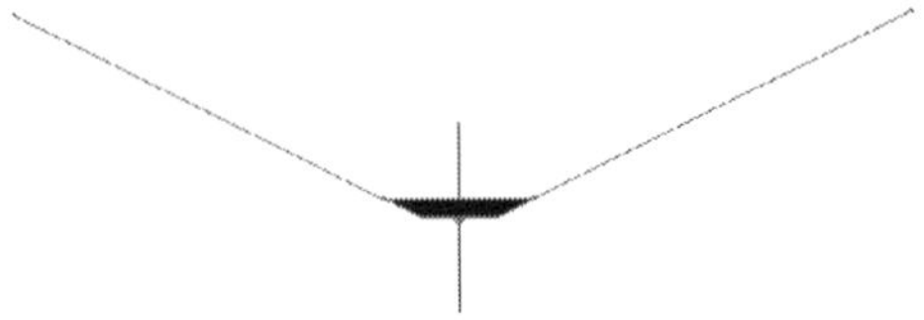

Figure 7-21 *The **PipeTrench** assembly*

Tip
*Select the subassembly after adding to the assembly and right-click to display the shortcut menu. Choose the **Subassembly Properties** from it. Next, choose the **Parameters** tab from the **Subassemblies Properties - Trench Pipe1** dialog box. Choose the **Subassembly help** button to display the **AutoCAD Civil 3D 2024 - English - Help** window. Study and understand the parameters and the subassembly diagram. Choose the **OK** button to close the dialog box. Next you can modify the input parameters in the **Properties** window.*

After understanding the input parameters of the **TrenchPipe1** subassembly, change the following input parameters in the **PROPERTIES** window:

Bedding Depth: **1.000'**
Sideslope: **2.00:1**
Width: **4.000'**

Chapter 8

Working with Corridors and Parcels

Learning Objectives

After completing this chapter, you will be able to:

- *Create simple and complex corridors*
- *Edit corridors*
- *Create corridor surfaces*
- *Add corridor boundaries*
- *Extract corridor solids and bodies*
- *Create parcels using objects*
- *View and edit parcel properties*

OVERVIEW

A corridor is a three-dimensional model of a design encompassing the horizontal, vertical, and the cross-sectional geometries. In other words, a corridor is a combination of an alignment that provides horizontal geometry, a profile that provides vertical geometry, and an assembly that provides cross-sectional geometry. These Civil 3D objects are combined together to create a dynamic three-dimensional corridor model.

Corridors for intersections, cul-de-sacs, and branching streams are complex models as they use multiple alignments and assemblies for their creation. Figure 8-1 shows the 2D view of a corridor in a plan view.

A corridor model displays different components such as points, point codes, links, link codes, subassemblies, feature lines, and surfaces that are using link codes and feature lines. These points and links are derived from subassemblies. Feature lines are the longitudinal lines that connect the identical point codes. These lines are created automatically when a corridor model is built. Each feature line is assigned the code similar to the point codes that the line connects to. Thus, feature line codes add a fourth dimension to the corridor model. Figure 8-2 shows the feature lines connecting the point codes and the 2D cross-section of assemblies placed after fixed intervals.

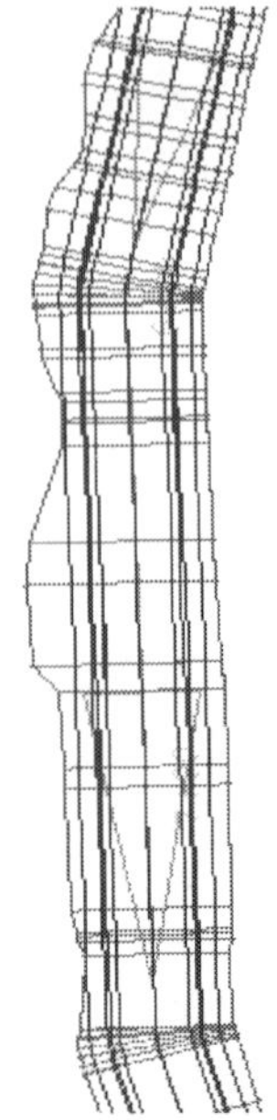

Figure 8-1 *2D view of a corridor model*

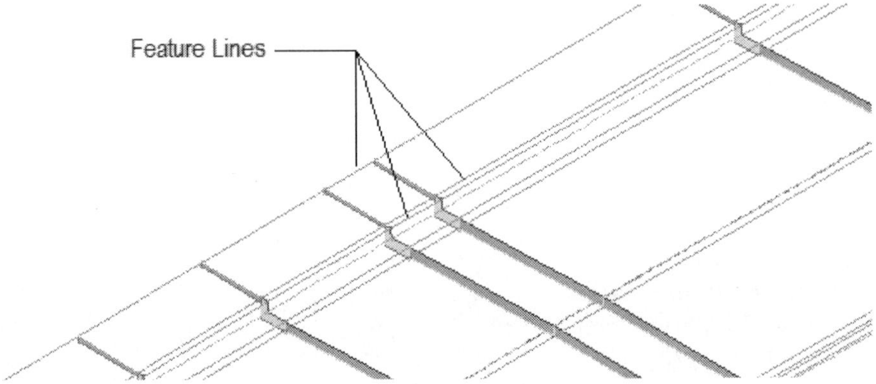

Figure 8-2 *Feature lines of the corridor*

CORRIDORS

A corridor can be a simple model created from a single alignment or a complex model created using multiple alignments. An alignment used for creating a corridor is called a baseline. Next, you will learn to create a corridor model using the **Create Corridor** interface in AutoCAD Civil 3D.

Creating Corridor

Ribbon:	Home > Create Design > Corridor
Command:	CREATECORRIDOR

The **Create Corridor** dialog box is used to specify the settings for creating a corridor. To create a corridor, choose the **Corridor** tool from the **Create Design** panel; the **Create Corridor** dialog box will be displayed, as shown in Figure 8-3.

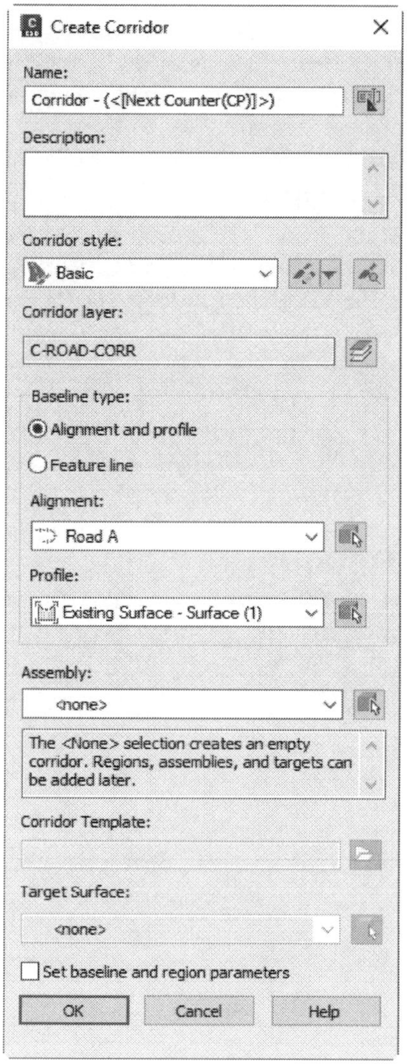

***Figure 8-3** The **Create Corridor** dialog box*

Note that to create a corridor in a drawing, you need to have a surface, an alignment, and a profile object in your drawing.

Corridor Properties

You can view and edit the corridor properties using the **Corridor Properties** dialog box. To invoke this dialog box, expand the **Corridors** node in the **Prospector** tab and then right-click on the existing corridor; a shortcut menu will be displayed. Choose the **Properties** option from the shortcut menu; the **Corridor Properties** dialog box will be displayed.

CORRIDOR SURFACE

A corridor surface is a component of corridor object in which the surface is created using the corridor subassembly point and link codes. You can create a corridor surface by using the feature line data, link data, or both in the **Surfaces** tab of the **Corridor Properties <corridor name>** dialog box. To invoke this dialog box, select the corridor from the drawing and right-click; a shortcut menu will be displayed. Choose the **Corridor Properties** from the shortcut menu; the **Corridor Properties <corridor name>** dialog box will be displayed. Next, choose the **Surfaces** tab from the dialog box and then use various options from it to create a corridor surface.

Creating a Corridor Surface

You can create a corridor surface that is dynamic in nature by using the options in the **Surfaces** tab of the **Corridor Properties** dialog box. To do so, choose the **Surface name template** button in the **Surfaces** tab of the **Corridor Properties <corridor name>** dialog box; the **Name Template** dialog box will be displayed. In the **Name Template** dialog box, enter the name of the corridor surface in the **Name** edit box and choose the **OK** button; the dialog box will be closed. Next, choose the **Create a corridor surface** button from the **Surfaces** tab in the **Corridor Properties <corridor name>** dialog box; a surface name will be added in the **Name** column of the table in the **Corridor Properties** dialog box. This surface will be empty, therefore, you need to add data to it to create a surface.

Adding the Surface Boundary

You can add boundaries to the corridor surface to prevent the formation of excess triangulation outside the corridor design. Figure 8-4 shows the corridor surface created by joining two codes and the extra surface triangulation formed outside the corridor surface.

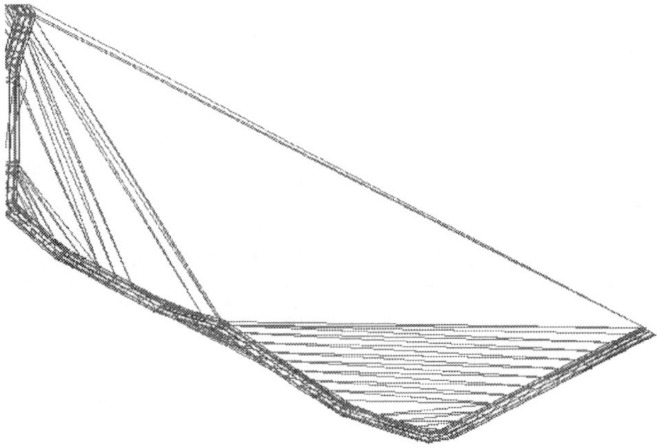

Figure 8-4 *The corridor surface created before adding the boundary*

To prevent this excess triangulation in the surface, add boundaries to the corridor surface. You can add surface boundaries using the **Boundaries** tab of the **Corridor Properties <corridor name>** dialog box. Invoke this dialog box by using the method explained in the earlier section. To add a boundary, choose the **Boundary** tab in the **Corridor Properties <corridor name>** dialog box and then right-click on the surface name in the **Name** column; a shortcut menu will be displayed, as shown in Figure 8-5.

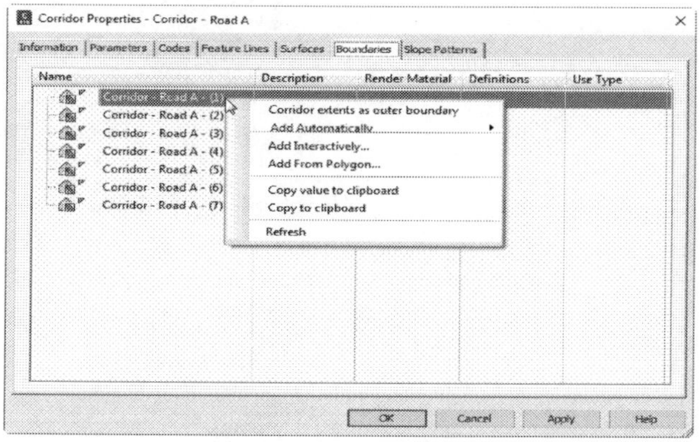

Figure 8-5 *The shortcut menu displaying various options*

Now, choose the required option from this shortcut menu to add the boundary to the corridor surface. Figure 8-6 shows the corridor surface created after adding the surface boundary.

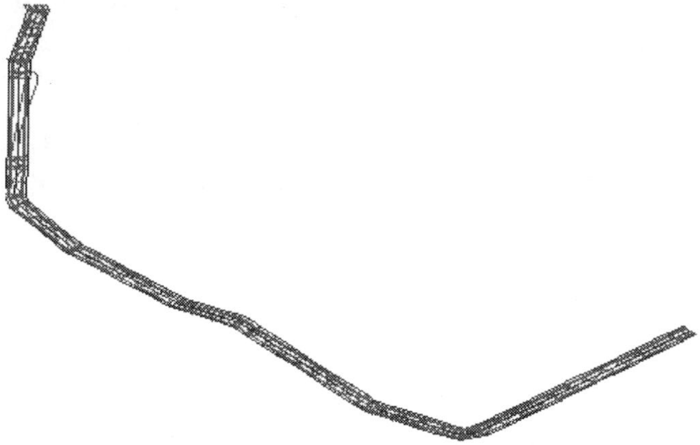

Figure 8-6 *Corridor surface created after adding the surface boundary*

Editing the Corridor Boundary

To edit the boundary added to the corridor surface, you need to select the corridor from the drawing and then right-click to display a shortcut menu. Choose **Corridor Properties** from the shortcut menu; the **Corridor Properties <corridor Name>** dialog box will be displayed. Choose the **Boundaries** tab from this dialog box and expand the required corridor surface node in the **Name** column.

To modify the name of the boundary, click on the existing name and then enter the desired name for the boundary in the edit box displayed. Similarly, click in the **Description**, **Render Material**, and **User Type** columns and modify the corresponding values. To view and modify the boundary definitions except for the one that is created from a polygon, click in the **Definitions** column; the **Corridor Boundary Definition** dialog box will be displayed, as shown in Figure 8-7.

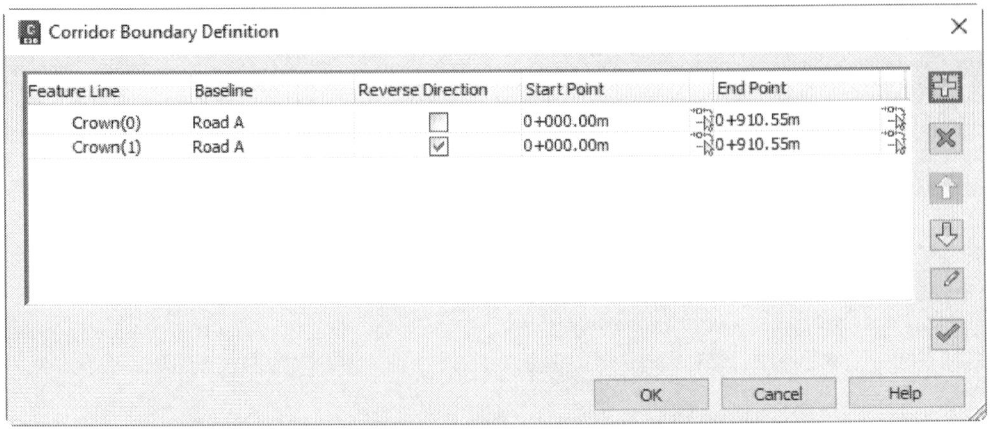

*Figure 8-7 The **Corridor Boundary Definition** dialog box*

Note

*You can change the render material of the boundary from the **Render Material** column of the **Boundary** tab provided the **Render Only** option is selected from the **User Type** column for the selected boundary.*

CREATING CIVIL 3D OBJECTS USING CORRIDOR

Civil 3D provides you with some tools that can be used to create objects from a corridor. These tools are discussed next.

Create Polyline From Corridor

Ribbon:	Home > Create Design > Create Polyline From Corridor
Command:	CREATEPOLYLINEFROMCORRIDOR

The **Create Polyline From Corridor** tool is used to create a polyline from the selected corridor feature line. To do so, choose the **Create Polyline From Corridor** tool from the **Create Design** panel; you will be prompted to select a corridor feature line. Select the required feature line from the corridor in the drawing; the polyline is created.

Note, due to the proximity of feature lines in the drawing, more than one feature line might be selected while selecting the feature line. In this case, the **Select a Feature Line** dialog box will be displayed. You can select the required feature line from this list of feature lines displayed in the dialog box and choose the **OK** button; the corridor feature line will be converted to a 3D polyline. Press ESC to end the command.

Create Feature Line from Corridor

Ribbon:	Home > Create Design > Feature Line drop-down > Create Feature Line from Corridor
Command:	FEATURELINESFROMCORRIDOR

You can create a grading feature line from the corridor and use it for feature line or projection grading. To do so, choose the **Create Feature Line from Corridor** tool from the **Create Design** panel; you will be prompted to select the corridor feature line. Select the corridor feature line in the drawing; the **Select a Feature Line** dialog box will be displayed. Select the required feature line from this dialog box and choose the **OK** button; the **Create Feature Line from Corridor** dialog box will be displayed. In this dialog box, specify the site, name, layer, and also choose the smoothing options for the grading feature line to be created. In this dialog box, the **Create dynamic link to the corridor** check box is selected by default. As a result, a dynamic link is created and maintained between the grading feature line and the corridor. Choose the **OK** button from this dialog box; the selected feature line will be converted into a grading line. Press ESC to end the command.

Create Alignment from Corridor

Ribbon:	Home > Create Design > Alignment drop-down > Create Alignment from Corridor
Command:	CREATEALIGNFROMCORRIDOR

An alignment can also be created from an existing corridor. This alignment is created along the horizontal path of the corridor feature line and is used to create profile views, special labeling, target alignments, and so on. To create an alignment from the corridor, choose the **Create Alignment from Corridor** tool from the **Create Design** panel; you will be prompted to select a corridor feature line. Select the corridor; the **Select a Feature Line** dialog box will be displayed. Select the required feature line from the dialog box and choose the **OK** button; the **Create Alignment from Objects** dialog box will be displayed. Enter the required information in the dialog box and choose the **OK** button. On doing so, the **Create Profile - Draw New** dialog box will be displayed. In this dialog box, specify the profile name in the **Name** edit box and choose the **OK** button; the **Create Profile - Draw New** dialog box will be closed. Press ESC to end the command. The required alignment will be created and listed in the **Alignments** node of the **Prospector** tab in the **TOOLSPACE** palette.

Create Profile from Corridor

Ribbon:	Home > Create Design > Profile drop-down > Create Profile from Corridor
Command:	CREATEPROFILEFROMCORRIDOR

To create a profile from the corridor, choose the **Create Profile from Corridor** tool from the **Create Design** panel; you will be prompted to select a corridor feature line. On doing so, the **Select a Feature Line** dialog box will be displayed. Select the required feature line from the dialog box and choose the **OK** button; the **Create Profile - Draw New** dialog box will be displayed. In this dialog box, set various parameters for the profile you want to create and then choose the **OK** button. Now, press ESC to end the command. Zoom in to view the profile view. The profile will be added to the **Profiles** sub node of the **Alignments** node in the **Prospector** tab.

Create COGO Points from Corridor

Ribbon:	Home> Create Ground Data > Points drop-down > Create COGO Points from Corridor
Command:	CREATEPOINTSFROMCORRIDOR

You can create the Civil 3D points using the corridor codes. To do so, choose the **Create COGO Points from Corridor** tool from the **Create Ground Data** panel; the **Create COGO Points** dialog box will be displayed, refer to Figure 8-8. Select the **For entire corridor range** radio button in this dialog box to create COGO points along the entire length of the corridor. You can also create the COGO points at a specified range along the corridor by selecting the **For user specified range** radio button in the **Create COGO Points** dialog box. Note that when you choose this radio button, the **Baseline start** and **Baseline end** edit boxes will be activated. Specify the start and the end station in the **Baseline start** and **Baseline end** edit boxes. The corridor codes that fall along the specified range of the baseline will be converted to COGO points.

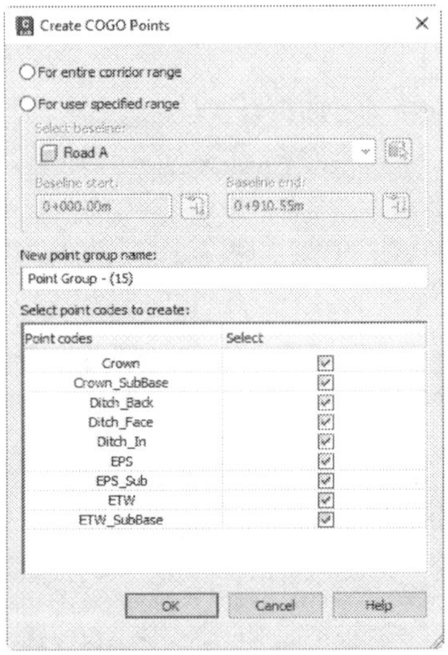

Figure 8-8 *Partial view of the* ***Create COGO Points*** *dialog box*

Now, in the **New point group name** edit box of the **Create COGO Points** dialog box, specify the name of the point group to which all COGO points will be added. Next, you need to select the point codes that will help in creating the points. To do so, you need to select the point codes from the **Select point codes to create** area. Select the check boxes in the **Select** column to select the corresponding point codes located in the **Point codes** column. Now, choose the **OK** button in the **Create COGO Points** dialog box; the dialog box will be closed and the COGO points will be generated for the specified corridor region using the selected point codes. The created points are added in the point group that you have specified and can be viewed under the **Point Groups** node of the **Prospector** tab in the **TOOLSPACE** palette.

Creating Surfaces from Corridor

Ribbon:	Home> Create Ground Data > Surfaces drop-down > Create Surface from Corridor
Command:	CORRIDOREXTRACTSURFACES

You can create a static surface object that is not linked to the corridor model using this tool. To create a corridor surface, choose the **Create Surface from Corridor** tool from the **Create Ground Data** panel; the **Create Corridor Surfaces** dialog box will be displayed. In this dialog box, a list of all the existing corridor surfaces is displayed. By default, all the check box under the **Select** column in this list are selected. Clear the check box for the surface which you do not intend to export. Next, click on the default value in the **Surface Style** column to select a surface style; the **Pick Corridor Surface Style** dialog box will be displayed. Select the required surface style from the drop-down list in this dialog box and choose the **OK** button; the selected surface style will be assigned to the corridor surface. Now, choose the **OK** button in the **Create Corridor Surfaces** dialog box; the dialog box will be closed and static surfaces are created for the surfaces corresponding to the selected check box. The names for these created surfaces are added to the **Surfaces** node of the **Prospector** tab.

CHECKING THE CORRIDOR VISIBILITY

Civil 3D provides various tools to check the visibility of the corridor model. These tools enable you to visually inspect your model along a specified path for the given section or entire length of the corridor. You can specify the path for checking the visibility by using Civil 3D objects such as the alignment, profile, feature line or 3D polyline.

The **Drive** tool and the **Check Sight Distance** tool, which are used to check corridor visibility are discussed next.

Drive

Ribbon:	Corridor (Contextual Tab) > Analyze > Drive
Command:	DRIVE

The **Drive** tool is simulation tool and is used to simulate driving along the corridor. To invoke the **Drive** tool, choose the corridor from the drawing; the **Corridor: <corridor name>** contextual tab is displayed. Choose the **Drive** tool from the **Analyze** panel of this tab; you will be prompted to select a corridor feature line. Select the required feature line object from the drawing. Civil 3D will process your request and display the simulated corridor in the drawing area. Figure 8-9 shows the simulated corridor.

Note

*In case more than one feature line is selected, the **Select a Feature Line** dialog box will be displayed with a list of the selected feature lines. Select the required feature line object from the displayed list and then choose the **OK** button in the dialog box.*

Figure 8-9 *Partial view of a corridor simulation*

Tip
*You can also access the **Drive** tool by choosing the **Drive** button in the **Design** panel of the **Analyze** tab.*

Sight Distance

Ribbon: Corridor (Contextual Tab) > Analyze > Sight Distance
Command: SIGHTDISTANCECHECK

This tool is used to analyze the sight distance along the corridor. To use this tool, you need to invoke the **Sight Distance Check** wizard. To do so, select the corridor from the drawing; the **Corridor: <corridor name>** contextual tab is displayed in the ribbon. In this tab, choose the **Sight Distance** tool from the **Analyze** panel; the **General** page of the **Sight Distance Check** wizard will be displayed, refer to the Figure 8-10. Alternatively, you can invoke the **Sight Distance Check** wizard by choosing the **Check Sight Distance** tool from the **Analyze > Design >Visibility check** drop-down.

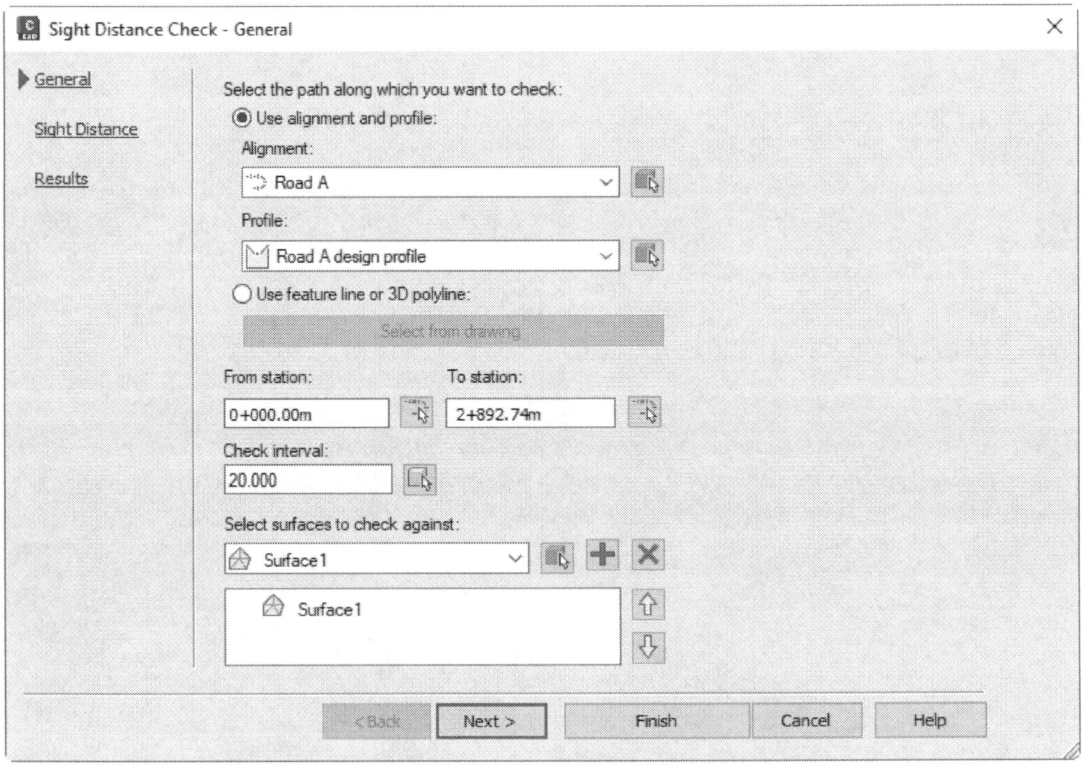

*Figure 8-10 The **General** page of the **Sight Distance Check** wizard*

INTRODUCTION TO PARCELS

In Civil 3D, parcels are objects which acts as a topology in a site. The entire project site will always be covered accurately with this topology. A site boundary is the outermost polygon containing interior land blocks and the smaller division of these blocks are referred as parcel objects. A site acts as a bucket for the parcel objects. Sites and parcels are closely related and they maintain active relationships with their boundaries and alignments. A site boundary can be divided into smaller parcels by inserting alignments within its boundary. For example, when you create an alignment in a site, the site will be divided into two parcels with the alignment acting as the dividing line. Also, if the alignment is deleted, the parcels combine and return to the original conditions.

In AutoCAD Civil 3D, parcels comprises series of points, lines, and curves and are always represented by closed areas with a label at the center or the centroid. A Civil 3D drawing may contain multiple sites and sites may contain multiple parcels. Therefore, you usually start with a site boundary and then divide the site into smaller parcels.

CREATING PARCELS

You can create parcels by using the existing objects in your drawing such as lines, curves, polylines or by using the **Parcel Layout Tools** toolbar. You can also create right of way parcel using the **Create Right of Way** tool.

Once a parcel is created, it is added to the **Parcel** sub-node under the **Sites** node in the **Prospector** tab of the **TOOLSPACE** palette. Various methods to create parcels are:

Creating Parcels using Polylines

Ribbon:	Home > Create Design > Parcel drop-down > Create Parcel from Objects
Command:	CREATEPARCELFROMOBJECTS

The simplest and most common method of creating a parcel is by using the AutoCAD drawing objects. Using this method, you can automatically convert the existing line work into parcels or parcel segments. Before using this tool, you require lines, arcs, or polylines to be present in your drawing and also all the objects should be closed polygons.

To convert the drawing objects into parcel objects, choose the **Create Parcel from Objects** tool from the **Create Design** panel of the **Home** tab; you will be prompted to select a line, an arc, or a polyline object. In the drawing, select the object you have drawn. Also, you can enter **X** in the command line to select the polyline from the Xref drawing. On selecting the required object from the drawing or from the Xref drawing, press ENTER; the **Create Parcels - From Objects** dialog box will be displayed, as shown in Figure 8-11. This dialog box is used to specify the information about the parcel. The options in this dialog box are discussed next.

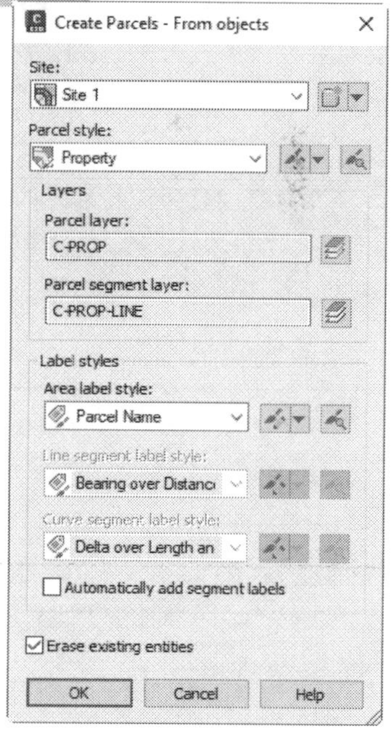

*Figure 8-11 The **Create Parcels - From objects** dialog box*

Creating Parcels Using the Parcels Layout Tools

Ribbon: Home > Create Design > Parcel drop-down > Parcel Creation Tools
Command: CREATEPARCELBYLAYOUT

AutoCAD Civil 3D allows you to create a parcel based on some complex designs by using the parcel layout tools. These tools can be accessed from the **Parcel Layout Tools** toolbar. These tools enable you to create new integrated designs for parcels.

Creating a parcel using the parcel layout tools is a two-step process. The first step includes the selecting of required parcel layout tool which you will use to draft the parcel from the **Parcel Layout Tools** toolbar. The second step involves naming the parcel, specifying the site, parcel layer, and parcel style.

To create a parcel using the parcel layout tools, choose **Parcel Creation Tools** from the **Create Design** panel; the **Parcel Layout Tools** toolbar will be displayed, as shown in Figure 8-12, and you will be prompted to select the required tool from the toolbar.

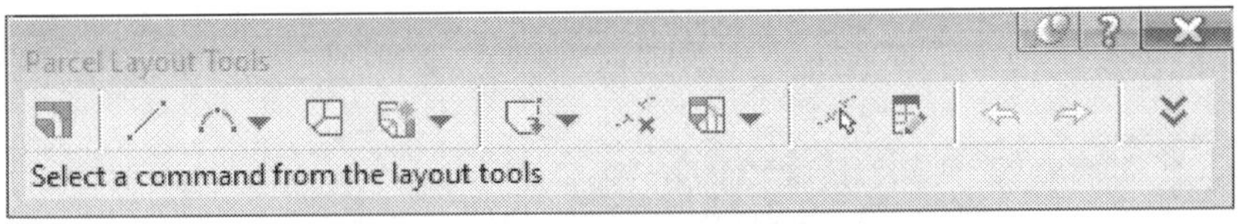

*Figure 8-12 The **Parcel Layout Tools** toolbar*

PARCELS LAYOUT TOOLS TOOLBAR

The tools in the **Parcel Layout Tools** toolbar are used to design a parcel object. Figure 8-13 shows the **Parcel Layout Tools** toolbar.

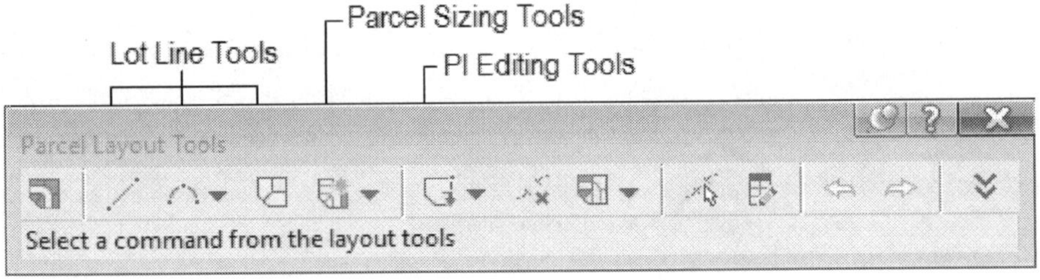

*Figure 8-13 The **Parcel Layout Tools** toolbar*

PARCEL PROPERTIES

You can view and edit the properties of a single parcel or a group of parcels. In both the cases, you can change the parcel style, area label style, parcel number, and address. An additional property is included in site parcel properties which specifies the display order of parcels within the site, including the site parcel itself. To edit parcel properties, select and right-click on a parcel area label in the drawing; a shortcut menu will be displayed. Choose the **Parcel**

Properties option; the **Parcel Properties - <Parcel Name>** dialog box with **Information** tab as default will be displayed as shown in Figure 8-14.

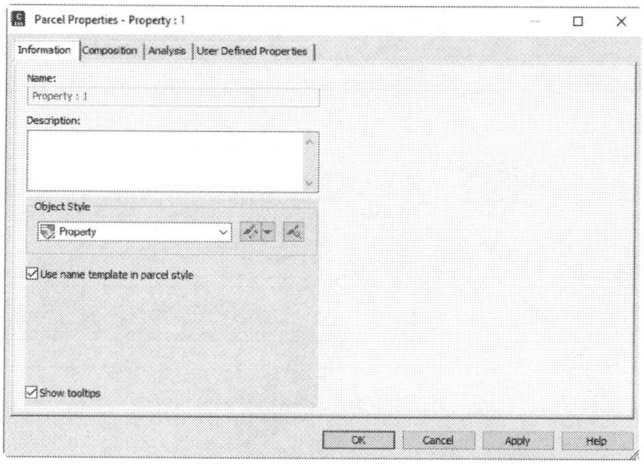

***Figure 8-14** Partial view of the **Parcel Properties - Property : 1** dialog box*

TUTORIAL

Before starting the tutorial, you need to download and save the tutorial files on your computer. To do so, follow the steps given below:

1. Log on to *www.cadsofttech.com* and browse to *Books > Civil/GIS Books > AutoCAD Civil 3D > AutoCAD Civil 3D 2024 for Novices*. Next, select *c08_c3d_2024_tut.zip* file from the **Tutorial Files** drop-down list. Choose the **Download** button to download the data file.

2. Now, save and extract the downloaded folder to the following location:

   C:\c3d_2024

Note
*While opening the tutorial file, the **PANORAMA** window may appear with an error message. Close this window to proceed further.*

Tutorial 1 — Creating a Corridor

In this tutorial, you will open the *c08_c3d_2024_tut* file that you downloaded from the Cadsofttech and then edit the corridor to create multiple regions and assign a new assembly (for the new region) to create a rest bay, as shown in Figure 8-15. **(Expected time: 30 min)**

The following steps are required to complete this tutorial:

a. Open the file.
b. Create a corridor.
c. View the corridor.

d. Create regions.
e. Save the file.

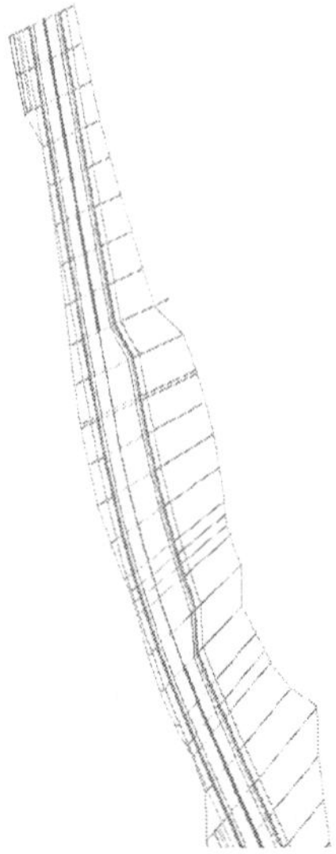

***Figure 8-15** Partial view of the Corridor showing the added rest bay*

Opening the File

1. Open the *c08_c3d_2024_tut.zip* file that you downloaded from *https://www.cadsofttech.com* and saved at *C:\c3d_2024*.

2. Choose **Open** from the **Application Menu**; the **Select File** dialog box is displayed.

3. In the dialog box, browse to the location *C:\c3d_2024\c08_c3d_2024_tut.*

4. Select the file *c08_c3d_2024_tut01.dwg* and then choose the **Open** button to open the file. Ignore the error message if displayed.

Creating a Corridor

In this section of the tutorial, you will create a corridor model using proposed road alignment, proposed profile, assembly, and existing ground surface.

1. Choose the **Corridor** tool from the **Create Design** panel of the **Home** tab; the **Create Corridor** dialog box is displayed.

2. Enter **Road Corridor** in the **Name** edit box.

3. Ensure that the **Proposed Road** option is selected in the **Alignment** drop-down list. Next, select the **Road Profile** option from the **Profile** drop-down list.

4. Select the **Road Assembly** and **Existing Ground** options from the **Assembly** and **Target Surface** drop-down lists, respectively.

5. Next, clear the **Set baseline and region parameters** check box and then choose the **OK** button; the **Create Corridor** dialog box is closed and Civil 3D starts executing the corridor creation process.

On the completion of the process, the **Road Corridor** is added to the **Corridors** node of the **Prospector** tab in the **TOOLSPACE**. Ignore the messages in the **PANORAMA** window if displayed and close the window.

Viewing the Corridor

1. Select and right-click on **Road Corridor** from the **Corridors** node in the **TOOLSPACE**; a shortcut menu is displayed.

2. Choose the **Select** option from the shortcut menu; the corridor is selected in the drawing window.

3. With the corridor selected, right-click in the drawing area; a shortcut menu is displayed.

4. Choose the **Isolate Objects** option from the shortcut menu; a cascading menu is displayed. Choose the **Isolate Selected Objects** option from the cascading menu, as shown in Figure 8-16; objects other than the selected corridor get hidden.

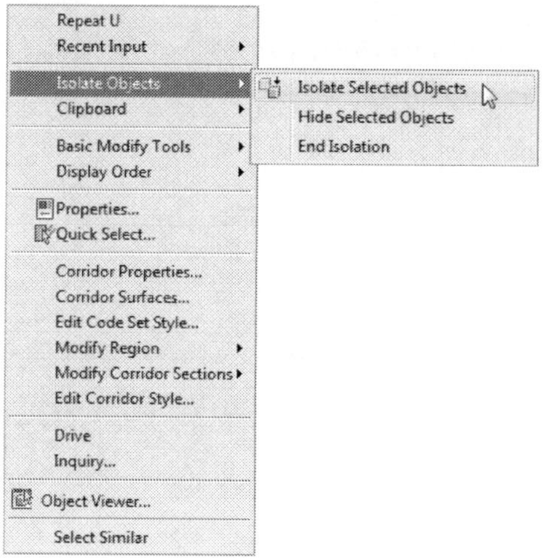

Figure 8-16 *Choosing the **Isolate Selected Objects** option*

5. Enter **ZE** in the command bar; the corridor in the drawing area zoom in, refer to Figure 8-17.

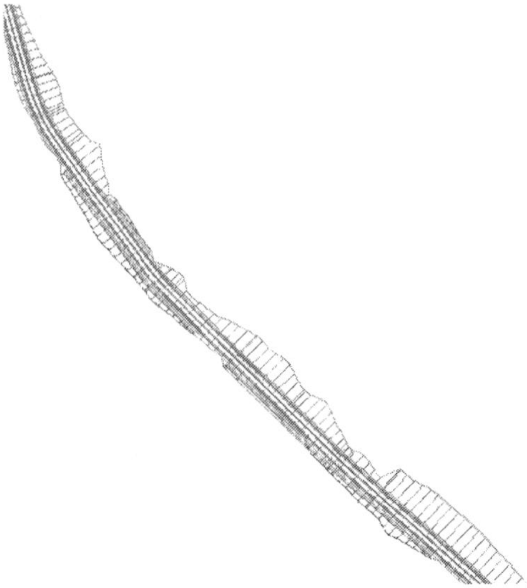

Figure 8-17 *Full view of the isolated **Road Corridor** object*

Tip

*You can also view the corridor model in the **Object Viewer** window. To do so, select the corridor model in the drawing and right-click; a shortcut menu will be displayed. Choose the **Object Viewer** option from the shortcut menu; the corridor model is displayed in the new **Object Viewer** window, as shown in Figure 8-18.*

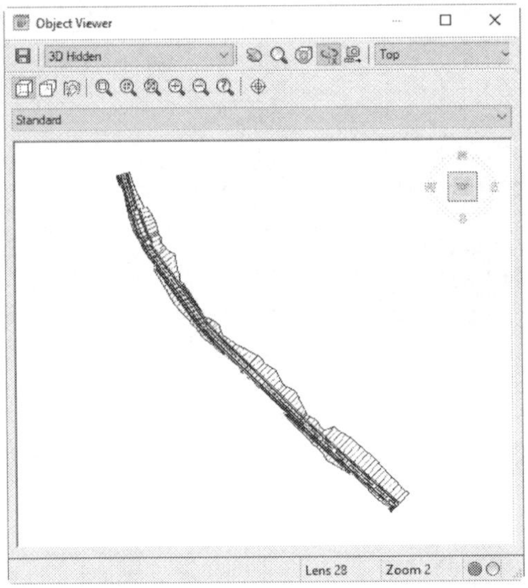

Figure 8-18 *The corridor model displayed in the **Object Viewer** window*

6. Right-click in the drawing; a shortcut menu is displayed. Choose **Isolate Object** from the shortcut menu; a cascading menu is displayed. Choose **End Object Isolation** from the cascading menu; the hidden objects are displayed again in the drawing area.

Creating Regions

1. Expand the **Corridors** node in the **Prospector** tab of the **TOOLSPACE**. Right-click on the **Road Corridor**; a shortcut menu is displayed. Choose **Properties** from the shortcut menu; the **Corridor Properties - Road Corridor** dialog box is displayed.

2. Choose the **Parameters** tab in this dialog box. Expand the **BL - Proposed Road - (1)** node in the **Name** column, if not already expanded. Right-click on **RG - Road Assembly - (1)** under the baseline **BL - Proposed Road - (1)** node; a shortcut menu is displayed.

3. Choose the **Split Region** option from the shortcut menu, as shown in Figure 8-19; the **Corridor Properties - Road Corridor** dialog box is closed and you are prompted to select a station along the alignment.

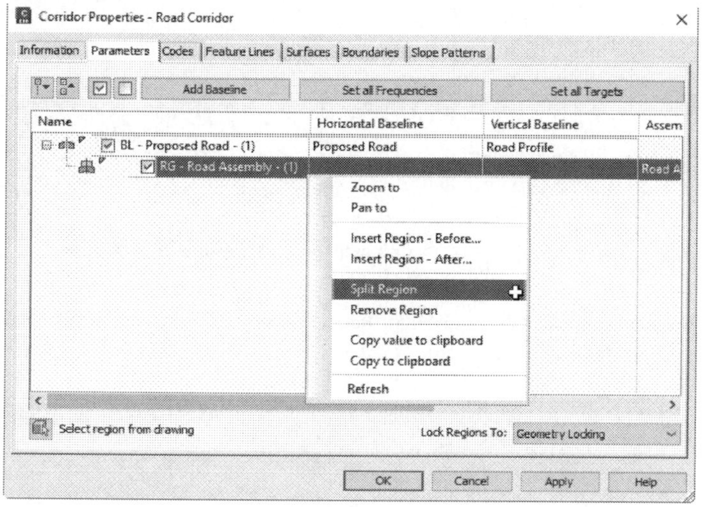

Figure 8-19 *Choosing the* ***Split Region*** *option from the shortcut menu*

4. Select a point along the alignment at station 2+00.00; a region is created at the selected station and you are prompted to specify another station.

5. Choose a point along the alignment at station **4+00.00**, as shown in Figure 8-20; a region is created at the selected station and you are prompted to specify another station.

Figure 8-20 *Specifying a point along the alignment at station* ***4+00.00***

6. Press ENTER to exit region creation; the **Corridor Properties - Road Corridor** dialog box is displayed again, refer to Figure 8-21.

Tip
*In case the corridor regions are not created at exact station values mentioned in the steps then edit the **Start Station** and **End Station** values manually for the regions in the **Corridor Properties** dialog box.*

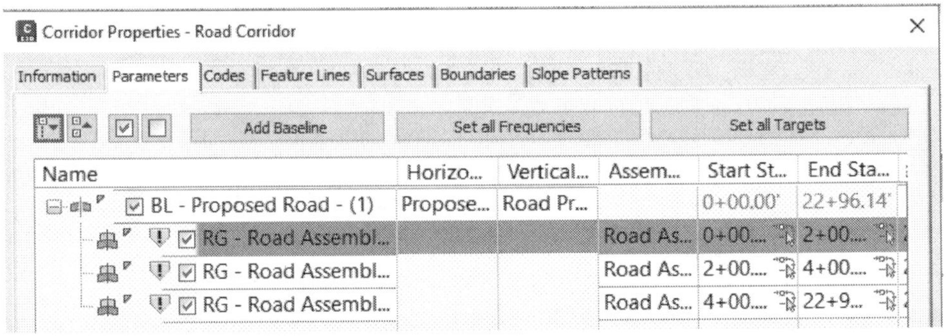

Figure 8-21 *Partial view of the **Corridor Properties - Road Corridor** dialog box with created regions*

7. Click on the **Name** cell of the newly created region **RG - Road Assembly - (1) (1)**; an edit box is displayed. Rename the region as **RG - Road Assembly - Rest Bay** in this edit box.

8. Next, click in the cell of the **Assembly** column corresponding to the **RG - Road Assembly - Rest Bay**; the **Edit Corridor Region** dialog box is displayed.

9. In the **Edit Corridor Region** dialog box, select the **Road Assembly- Rest bay** option from the **Assembly** drop-down list. Next, choose the **OK** button; the **Edit Corridor Region** dialog box is closed. The **Road Assembly-Rest bay** name is displayed in the **Assembly** column for the **RG - Road Assembly - Rest Bay** region in the **Corridor Properties - Road Corridor** dialog box. Ignore the message in the **PANORAMA** window if displayed.

10. Next, choose the **OK** button; the **Corridor Properties - Rebuild** message box is displayed, as shown in Figure 8-22.

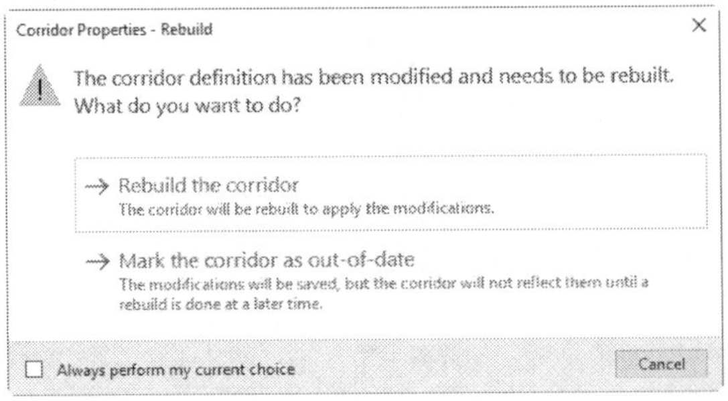

Figure 8-22 *The **Corridor Properties - Rebuild** message box*

11. Choose the **Rebuild the corridor** option from the message box; the message box is closed and the corridor model is rebuilt.

After completing the rebuild process, follow the steps explained in the **Viewing the Corridor** section of this tutorial to view the corridor model.

12. Zoom to the corridor along station 2+00.00 to 4+00.00. Notice the rest bay is added to the corridor, refer to Figure 8-23.

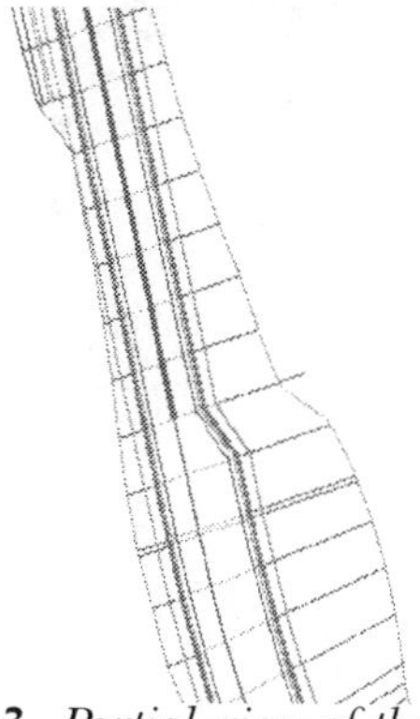

Figure 8-23 *Partial view of the Corridor showing the added rest bay*

Saving the File

1. Choose **Save As** from the **Application Menu**; the **Save Drawing As** dialog box is displayed.
2. In this dialog box, browse to the following location:

   C:\c3d_2024\c08_c3d_2024_tut

3. In the **File name** edit box, enter **c08_tut01a**.
4. Choose the **Save** button; the file is saved with the name *c08_tut01a.dwg* at the specified location.

EXERCISE

1. Log on to *www.cadsofttech.com* and browse to *Books > Civil/GIS Books > AutoCAD Civil 3D > AutoCAD Civil 3D 2024 for Novices*. Next, select *c08_c3d_2024_exer.zip* file from the **Exercise Files** drop-down list. Choose the **Download** button to download the data file.
2. Now, save and extract the downloaded folder to the following location:

   C:\c3d_2024

Note

*While opening the exercise files, the **PANORAMA** window may appear with an error message. Close this window to proceed further.*

Exercise 1

Create a corridor, as shown in Figure 8-24, using the *c08_c3d_2024_ex01.dwg* file. Create corridor for Region 2 and hide Region 1 of the corridor. Use the following parameters:

(Expected time: 30 min)

Corridor name:	**Canal 2**
Surface Style for Region 2:	**Canal Surface**
Assembly:	**Canal 2**
End Station for Region 1:	**135+00.00'**
Start Station for Region 2:	**195+00.00'**

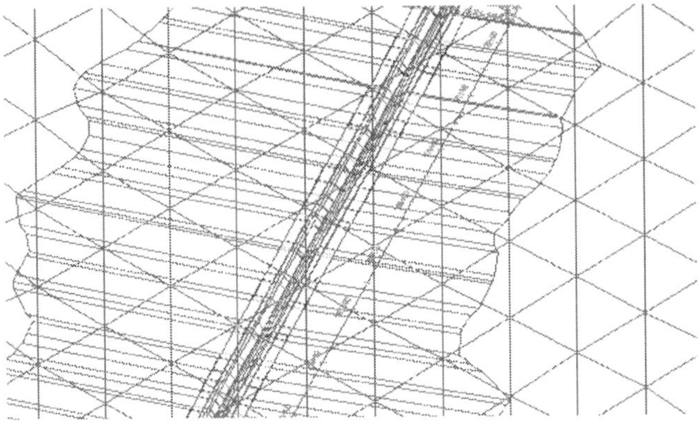

Figure 8-24 *The corridor model*

Chapter 9

Sample Lines, Sections, and Quantity Takeoffs

Learning Objectives

After completing this chapter, you will be able to:

- *Know about sample lines*
- *Create sample lines*
- *Edit sample lines*
- *Create section views*
- *Understand quantity takeoffs*

OVERVIEW

In AutoCAD Civil 3D, a cross section provides a graphical view of the terrain or surface elevations along a linear cut feature at certain angle across a horizontal alignment. To create a cross section, you need at least two Civil 3D objects such as an alignment and a surface. Thus, a cross section consists of elevation data along any linear feature. The major components of a section are vertices or points at which one grade line ends and another starts. The other component of the section is the line segment that represents the surface grade between two vertices or grade break points.

In Civil 3D, you can create sections from surfaces, corridors, corridor surfaces, and pipe networks along a linear object.

A section is created along the sample line and is displayed using a section view. These section views are graphical representation of the elevation data or an assembly (in case of corridor sections) at a given station. Figure 9-1 shows a typical cross section view of a road corridor.

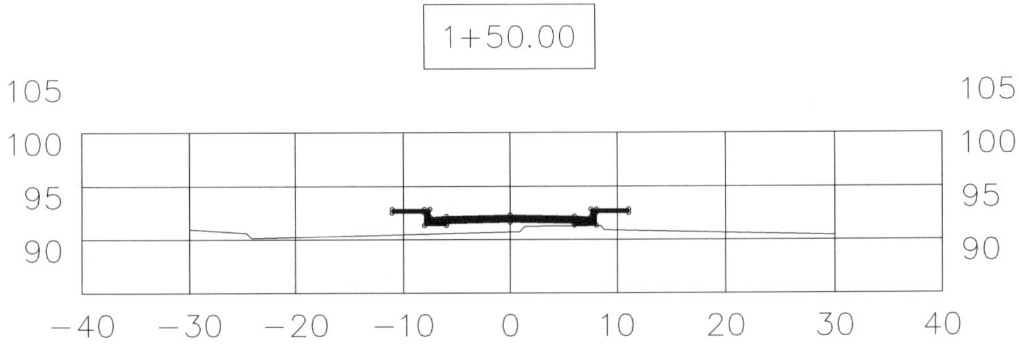

***Figure 9-1** A typical section view of a road corridor*

SAMPLE LINES

A sample line is a linear object that is used for creating a cross section across an alignment or a corridor. A sample line represents the direction along which the section is cut. A sample line consists of a linear feature and sample line label. A group of sample lines is called sample line group. You can create multiple groups of sample lines for a single alignment. The sample line group controls the display of sample lines, sections, and section views. The sample line group is available in the parent alignment node in the **Prospector** tab of the **TOOLSPACE** palette.

Sample lines are created using the **Sample Lines** tool. Before you start creating sample lines, you need to have an alignment. It is better to create a corridor and corridor surfaces before creating sample lines because corridor can provide you more information such as materials used and surfaces with cut and fill areas.

Creating Sample Lines

Ribbon: Home > Profile & Section Views > Sample Lines
Command: CREATESAMPLELINES

To create sample lines, choose the **Sample Lines** tool from the **Profile & Section Views** panel; you will be prompted to select an alignment from the drawing. Alternatively, press ENTER to select alignment from the list; the **Select Alignment** dialog box will be displayed. Select the required alignment from this dialog box and then choose the **OK** button; the dialog box will be closed and the **Sample Line Tools** toolbar will be displayed, as shown in Figure 9-2. In case you are creating a sample line for the first time for a given alignment, the **Create Sample Line Group** dialog box will also be displayed along with the **Simple Line Tools** toolbar. Options in this dialog box are discussed later in this chapter.

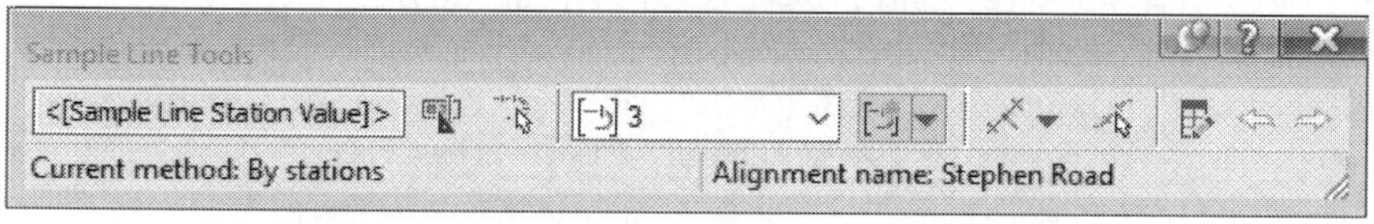

Figure 9-2 *The **Sample Line Tools** toolbar*

In the **Create Sample Line Group** dialog box specify the required parameters and choose the **OK** button; the dialog box will be closed and you will be prompted to specify a station. You can specify stations by using different options available in the **Sample Line Tools** toolbar. The tools in the **Sample Line Tools** toolbar are used to create and edit sample lines along the alignment.

Sample Line Properties

To view and edit the properties of a sample line, select the required sample line from the drawing and right-click; a shortcut menu will be displayed. Choose the **Sample Line Properties** option from the shortcut menu; again select the required sample line from the drawing; the **Sample Line Properties** dialog box will be displayed, as shown in Figure 9-3.

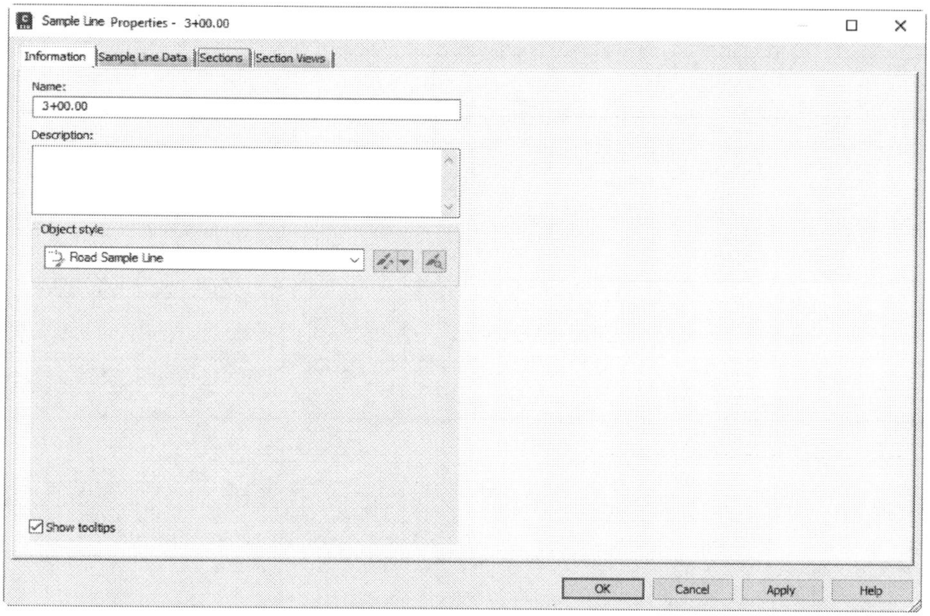

Figure 9-3 *The **Sample Line Properties - 3+00.00** dialog box*

Alternatively, to invoke the **Sample Line Properties** dialog box, expand the **Sample Line Groups** sub-node with which the required sample line is associated. Right-click on the required sample line; a shortcut menu is displayed. Choose the **Properties** option from the shortcut menu; the **Sample Line Properties** dialog box will be displayed.

Sample Line Group Properties

To view the sample line group properties, expand the **Alignments** node from the **Prospector** tab of the **TOOLSPACE** palette. Next, expand the required alignment node with which the sample line group is associated. Now, expand the **Sample Line Groups** sub-node under the alignment node; the existing sample line groups will be listed in this node. Right-click on the required sample line group; a shortcut menu will be displayed. Choose the **Properties** option from the shortcut menu; the **Sample Line Group Properties** dialog box will be displayed.

Editing Sample Lines

Ribbon: Sample Line > Modify > Edit Sample Line
Command: EDITSAMPLELINE

To edit sample lines parameters, select the sample line from the drawing; the **Sample Line - <station number>** tab will be displayed. In this tab, choose the **Edit Sample Line** tool from the **Modify** panel; the **Edit Sample Line** dialog box will be displayed along with the **Sample Line Tools** toolbar. Also, you will be prompted to select a sample line in the command line. Select the required sample line; the **Edit Sample Line** dialog box will be populated with the selected sample line parameters. You can edit the required parameters either by using this dialog box or by using the **Sample Line Tools** toolbar, as explained earlier in this chapter.

To edit the sample line style, select the **Edit Sample Line Style** tool from the **Sample Line Properties** drop-down from the **Modify** panel of the **Sample Line** tab; the **Sample Line Style - <Sample Line Style Name>** dialog box is displayed. You can use this dialog box to edit the sample line style. Alternatively, you can invoke this dialog box from the **Settings** tab of the **TOOLSPACE** palette. To do so, in the **Settings** tab, expand **Sample Line > Sample Line Styles** sub-node. Next, select the required sample line and right-click on it; a shortcut menu will be displayed. Choose the **Edit** option from the shortcut menu, as shown in Figure 9-4 the **Sample Line Style** dialog box will be displayed.

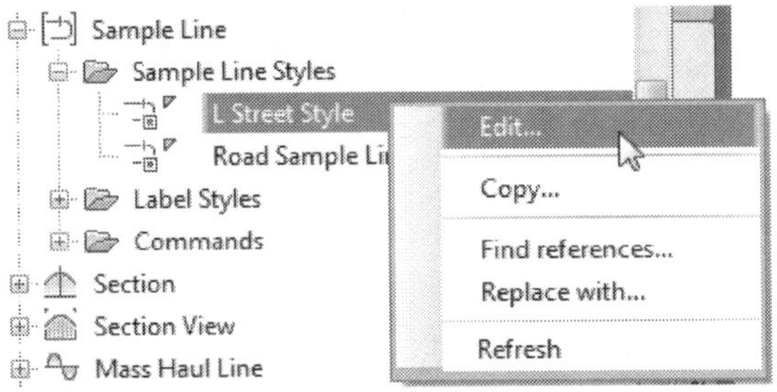

Figure 9-4 *Choosing the **Edit** option from the shortcut menu*

CREATING SECTION VIEWS

After defining a sample line group and creating the sample lines, you will create section views. In AutoCAD Civil 3D, you can create two types of section views, single section view and multiple section view. Section views are added to the collection of parent sample line in the **Prospector** tab. The processes to create different section views are discussed next.

Creating a Single Section View

Ribbon:	Home > Profile & Section Views > Section Views drop-down > Create Section View
Command:	CREATESECTIONVIEW

To create a single section view at a particular station, choose the **Create Section View** tool from the **Profile & Section Views** tab; the **General** page of the **Create Section View** wizard will be displayed, as shown in Figure 9-5.

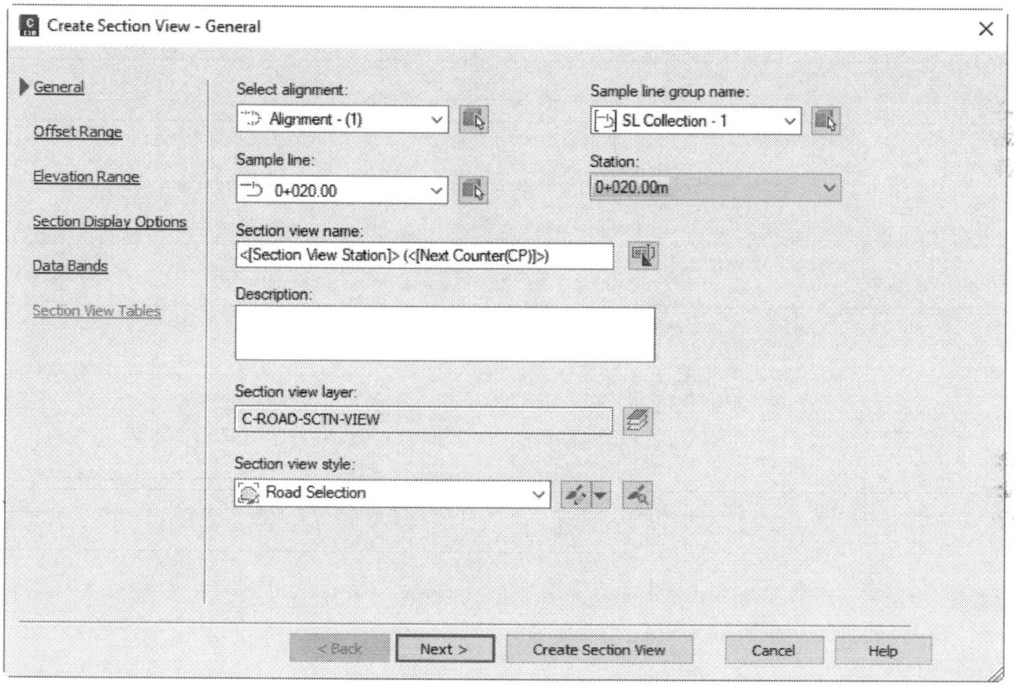

Figure 9-5 *The* ***General*** *page of the* ***Create Section View*** *wizard*

The **Create Section View** wizard is composed of various pages such as the **General**, **Offset Range**, **Elevation Range**, **Section Display Options**, **Data Bands**, and **Section View Tables**. Each page hosts a number of tools and options for setting the criteria and the parameters to display in the section view.

Creating Multiple Section Views

Ribbon: Home > Profile & Section Views > Section Views drop-down > Create Multiple Views
Command: CREATEMULTIPLESECTIONVIEW

To create multiple section views, choose the **Create Multiple Views** tool from the **Profile & Section Views** panel; the **General** page of the **Create Multiple Section Views** wizard will be displayed, as shown in Figure 9-6. This wizard is used to create multiple section views from the existing sections. The pages in this wizard are the same as those discussed in the **Create Section View** wizard.

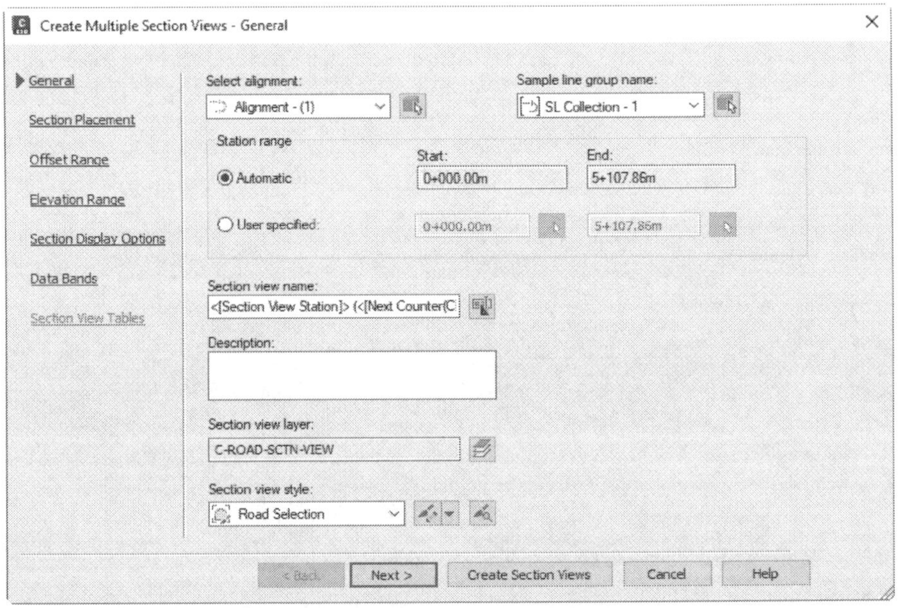

*Figure 9-6 The **General** page of the **Create Multiple Section Views** wizard*

To display the **Section Placement** page, choose the **Next** button in the **General** page of the **Create Multiple Section Views** wizard.

QUANTITY TAKEOFFS

Civil 3D provides you with two types of quantity takeoffs namely, earthworks and material volumes. You can generate the formatted reports of earthwork and material volumes. The Earthwork takeoff represents the amount of total material to be added to or removed from the existing site. This is done by comparing the existing surface with the datum of the assembly. The datum of the assembly is the bed over which different layers of materials (rubble, concrete, and so on) to build the road are laid. The main purpose of the earthwork calculation is to modify the design such that a balance is maintained between the total cut and fill material.

TUTORIAL

Before starting the tutorial, you need to download and save the tutorial files on your computer. To do so, follow the steps given below:

1. Log on to *www.cadsofttech.com* and browse to *Books > Civil/GIS Books > AutoCAD Civil 3D > AutoCAD Civil 3D 2024 for Novices*. Next, select *c09_c3d_2024_tut.zip* file from the **Tutorial Files** drop-down list. Choose the **Download** button to download the data file.

2. Now, save and extract the downloaded folder to the following location:

   C:\c3d_2024

Note

*While opening the tutorial file, the **PANORAMA** window may appear with an error message. Close this window to proceed further.*

Tutorial 1 — Section View

In this tutorial, you will open the *c09_c3d_2024_tut* file that you downloaded from the Cadsofttech and then create a sample line group by using the **From Corridor Stations** method and then create a section view at the selected station, as shown in Figure 9-7. **(Expected time: 1 hr)**

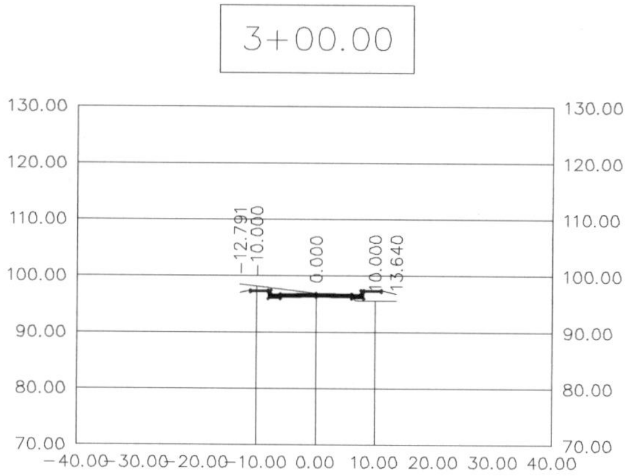

***Figure 9-7** The created section view at the selected station*

The following steps are required to complete this tutorial:

a. Open the file.
b. Create the sample line group.
c. Create the cross section at the selected station.
d. Edit the section label style.
e. Save the file.

Opening the File

1. Open the *c09_c3d_2024_tut.zip* file that you downloaded from *https://www.cadsofttech.com,* as explained in Chapter 2.

2. Choose **Open > Drawing** from the **Application Menu**; the **Select File** dialog box is displayed.

3. In the dialog box, browse to the location *C:\c3d_2024\c09_c3d_2024_tut.*

4. Select the file *c09_c3d_2024_tut01.dwg* and then choose the **Open** button to open the file.

Ignore any error messages if displayed in the **PANORAMA** window while opening the drawing file.

Creating a Sample Line Group

1. Choose the **Sample Lines** tool from the **Profile & Section Views** panel of the **Home** tab; you are prompted to select an alignment.

2. Press ENTER; the **Select Alignment** dialog box is displayed.

3. Select **Stephen Road** from the **Select Alignment** dialog box and then choose the **OK** button; the **Sample Line Tools** toolbar with the **Create Sample Line Group** dialog box is displayed.

4. In this dialog box, choose the **Click to edit name template** button on the right of the **Name** edit box; the **Name Template** dialog box is displayed. 

5. In the **Name Template** dialog box, enter **Stephen Collection** in the **Name** edit box and choose the **OK** button; the **Name Template** dialog box is closed.

6. Select the **Stephen Style** option from the **Sample line style** drop-down list in the **Create Sample Line Group** dialog box.

7. Next, select **SR Label Style** from the **Sample line label style** drop-down list and then choose the **OK** button; the **Create Sample Line Group** dialog box is closed.

8. Make sure that the **Stephen Collection** option is selected in the **Current sample line group** drop-down list of the **Sample Line Tools** toolbar and then choose the down-arrow next to the **Sample line creation methods** button; a drop-down list is displayed. 

9. Choose the **From corridor stations** tool from the drop-down list, as shown in Figure 9-8; the **Create Sample Lines - From Corridor Stations** dialog box is displayed.

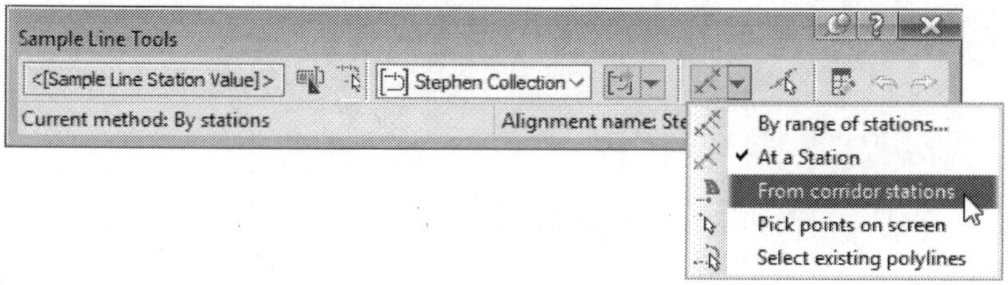

***Figure 9-8** Choosing the **From corridor stations** tool from the flyout*

10. In the **Station Range** category of this dialog box, set the value of the **To alignment end** property to **False**.

11. For the **End Station** property, enter **5+19.91'** in the **Value** column and then press ENTER.

12. In the **Left Swath Width** category, click in the value field of the **Width** property; a button is displayed on the right of the **Value** column.

13. Choose the button; the **Create Sample Line - From Corridor Stations** dialog box is closed and you are prompted to specify the distance.

14. Zoom in the first station at the start of the corridor and click at the top center of the alignment, as shown in Figure 9-9; you are prompted to specify the second point.

15. Drag the cursor to the left, as shown in Figure 9-10, and then click on the screen to specify the left swath width. On doing so, the **Create Sample Lines - From Corridor Stations** dialog box is displayed again and the swath width is displayed in the **Value** column.

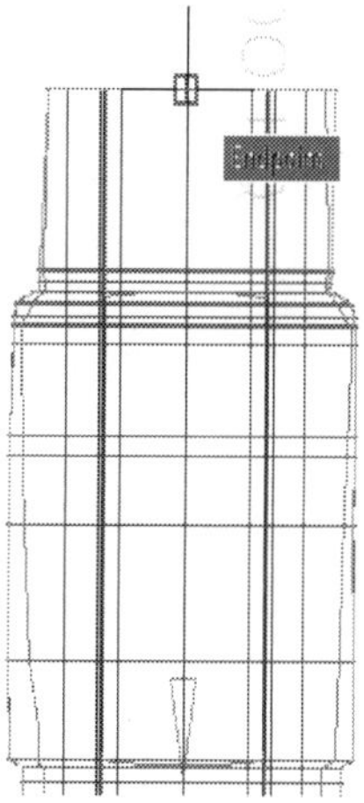

***Figure 9-9** Clicking at the top center of the alignment*

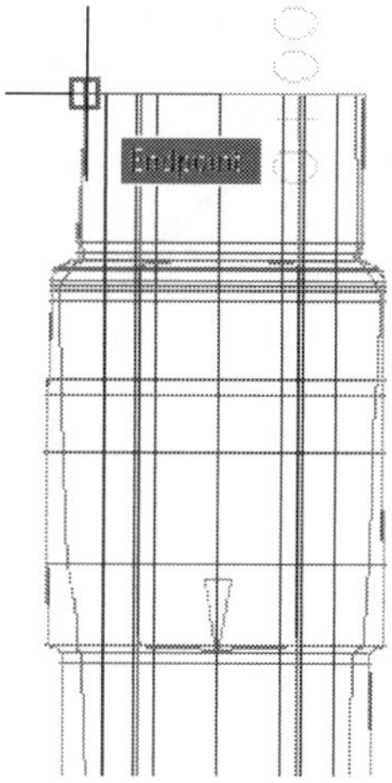

***Figure 9-10** Dragging the cursor on the left to specify the swath width*

16. Now, set the value of the swath width in the **Right Swath Width** edit box to **13.640**.

17. Choose the **OK** button; the **Create Sample Lines - From Corridor Stations** dialog box is closed.

18. Next, press ESC to exit the command.

19. Close the **PANORAMA** window, if it is displayed; sample lines are created from corridor stations, as shown in Figure 9-11.

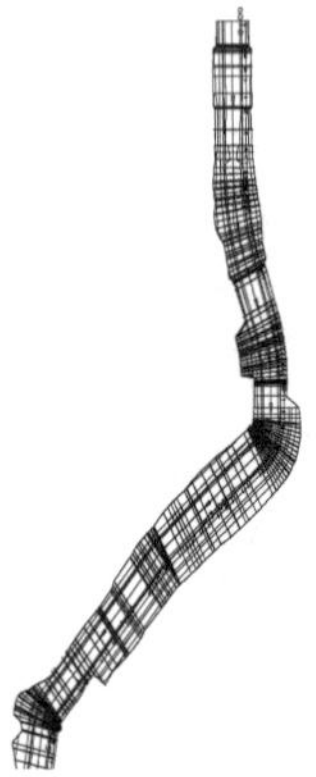

Figure 9-11 *Sample lines created from corridor stations*

Creating the Section View

1. Choose the **Create Section View** tool from **Home > Profile & Section Views > Section Views** drop-down; the **General** page of the **Create Section View** wizard is displayed.

2. Choose the **Click to edit name template** button, next to the **Section view name** edit box on this page; the **Name Template** dialog box is displayed.

3. In the **Name Template** dialog box, enter **Stephen Section** in the **Name** edit box and then choose the **OK** button; the dialog box is closed.

4. Next, choose the down-arrow next to the **Section view style** drop-down list; a flyout is displayed.

5. Choose the **Create New** option from the flyout; the **Section View Style - New Section View Style** dialog box is displayed.

6. Choose the **OK** button to accept the default settings in this dialog box.

7. Next, from the **Section view style** drop-down list on the **General** page, select **New Section View Style**, if it is not selected by default.

8. Select the **3+00.0** option from the **Sample line** drop-down list.

9. Next, choose the **Section Display Options** option from the left panel of this wizard; the **Section Display Options** page is displayed.

10. In the **Select sections to draw** list box of the **Section Display Options** page, clear the check box in the **Draw** column for the **corridor 1 Top surface** row to exclude drawing of the top surface of the corridor in the section view.

11. Now, choose the **Create Section View** button; the **Create Section View** wizard is closed and you are prompted to specify the origin of the section view.

12. Click anywhere in the drawing; the section view is created, as shown in Figure 9-12. Press ESC to exit the command.

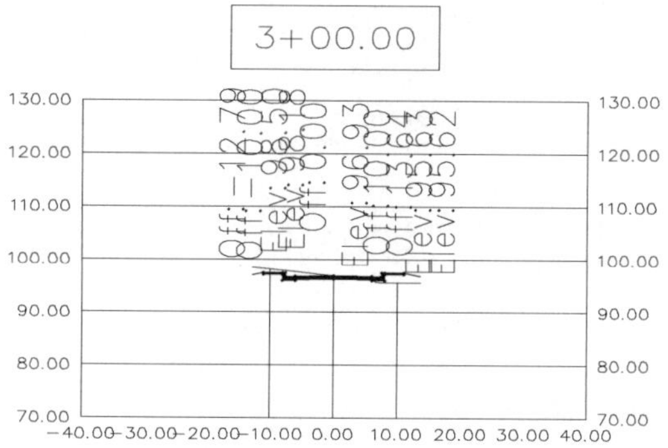

Figure 9-12 The section view created at alignment station **3+00.00**

Editing the Section Label Style

1. Select the section view from the drawing and right-click; a shortcut menu is displayed.

2. Choose **Section View Properties** option from the shortcut menu; the **Section View Properties - Stephen Section** dialog box is displayed,

3. Choose the **Sections** tab from this dialog box, refer to Figure 9-13.

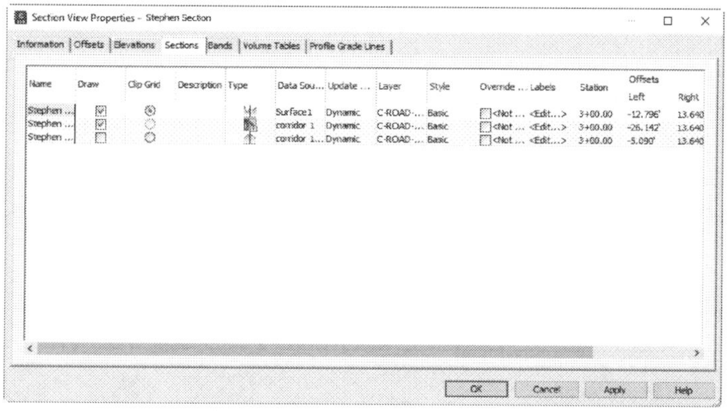

***Figure 9-13** Partial view of the options displayed in the **Sections** tab of the **Section View Properties - Stephen Section** dialog box*

4. Click on **<Edit>** in the **Labels** column corresponding to **Stephen Collection - 3+00.00 - Surface1** (first row); the **Section Labels - Stephen Collection - 3+00.00 - Surface 1(1093)** dialog box is displayed. Make sure that the **Major Offset** and **Basic** options are selected in the **Type** and **Section Major Offset Label Style** drop-down lists, respectively, of the dialog box.

5. Click on the icon displayed in the **Style** column; the **Pick Label Style** dialog box is displayed. 

6. Select **Stephen Major** from the drop-down list in the dialog box, as shown in Figure 9-14.

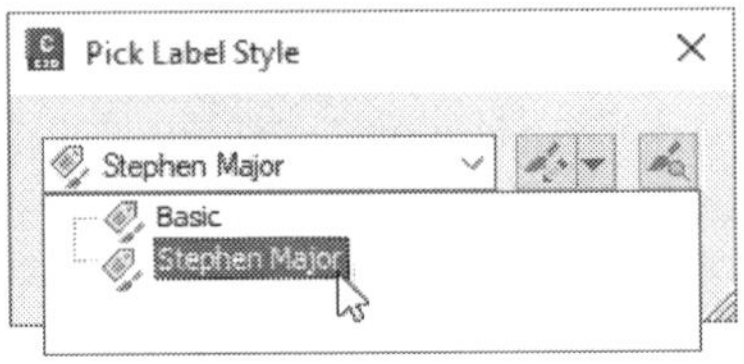

Figure 9-14 *Selecting the* ***Stephen Major*** *option from the drop-down list*

7. Choose the **OK** button; the **Pick Label Style** dialog box is closed.

8. Choose **OK** in the **Section Labels - Stephen Collection - 3+00.00 - Surface1(1093)** dialog box; the dialog box is closed and the **Section View Properties - Stephen Section** dialog box is displayed.

9. Now, choose the **OK** button in the **Section View Properties - Stephen Section** dialog box; the dialog box is closed and the section view is displayed, as shown in Figure 9-15. Similarly, you can create a section view at any station by selecting the required sample line.

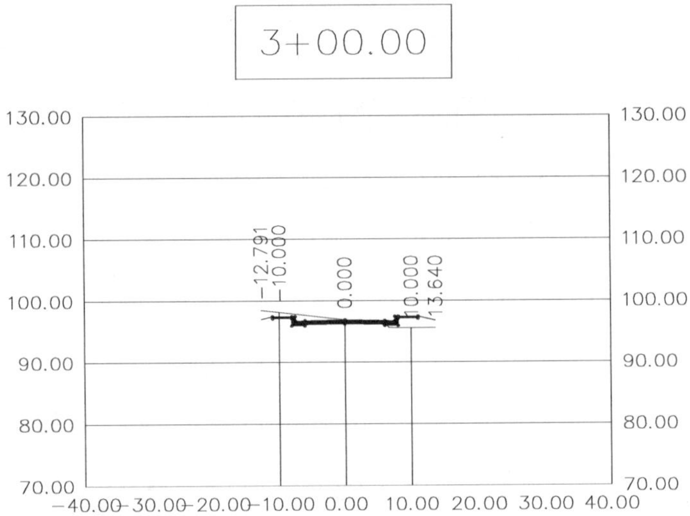

Figure 9-15 *The* ***Stephen Section*** *view created at station* ***3+00.00***

Saving the File

1. Choose **Save As** from the **Application Menu**; the **Save Drawing As** dialog box is displayed.
2. In this dialog box, browse to the following location:

   C:\c3d_2024\c09_c3d_2024_tut

3. In the file name edit box, enter **c09_tut01a.**
4. Choose the **Save** button; the file is saved with the name *c09_tut01a.dwg* at the specified location.

EXERCISE

1. Log on to *www.cadsofttech.com* and browse to *Books > Civil/GIS Books > AutoCAD Civil 3D > AutoCAD Civil 3D 2024 for Novices*. Next, select *c09_c3d_2024_exer.zip* file from the **Exercise Files** drop-down list. Choose the **Download** button to download the data file.
2. Now, save and extract the downloaded folder to the following location:

   C:\c3d_2024

Note
*While opening the exercise files, the **PANORAMA** window may appear with an error message. Close this window to proceed further.*

Exercise 1

Open the *c09_c3d_2024_ex01.dwg* file and create a group plot style and multiple section views, as shown in Figure 9-16. **(Expected time: 30 min)**

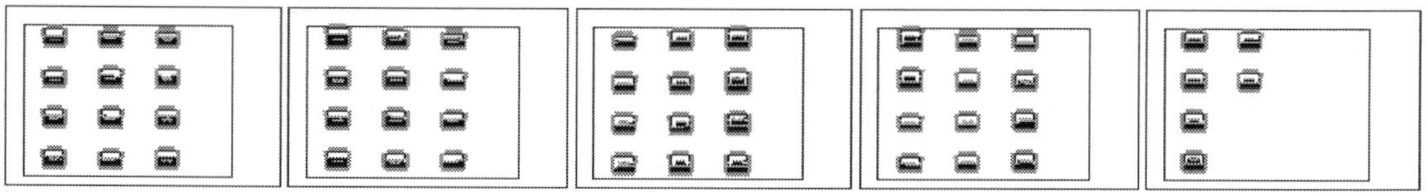

Figure 9-16 *Multiple section views*

Use the following settings for the group plot style:

- Maximum in a row: **7**
- Space between adjacent sections in a row and column: **2.0000"**
- Save the drawing as *c09_ex_01a.dwg*
- Accept the default values for all other options

This page is intentionally left blank

Chapter 10

Feature Lines and Grading

Learning Objectives

After completing this chapter, you will be able to:

- *Create feature lines*
- *Edit feature lines*
- *Understand grading objects and grading groups*
- *Understand grading criteria sets*
- *Create grading and grading styles*

OVERVIEW

A feature line is a Civil 3D object that is used to represent important features in a drawing such as ridge line and pavement edges. Feature lines are essentially 3D polylines with some additional properties and data that a polyline does not possess. In the grading process, the feature lines are used to create a closed or open geometric figure (Footprint) to which the grading criteria can be applied.

Grading is a process of modeling finished ground. In this process, you add surfaces to the site at different elevations and slopes to achieve the required level of surface. This process is carried out throughout the design process. For example, at the initial stage of site development, rough grading is done by the planner to verify the proposed site plans. After verifying the site plan, the rough grading data is used to estimate the required design road grades by creating the grading objects and groups, baseline geometry, and surfaces. Finally, the grading of finished ground is done to ensure that the maximum and minimum slope criteria are fulfilled.

Grading objects and feature lines are two main prerequisites for completing a grading process. While the grading groups are created using the grading layout tools, the features lines are created using feature lines tool. Before you proceed further with the grading process, you need to understand the concept of feature line object as well as different tools associated with it.

FEATURE LINES

Feature lines can be created by using various feature line tools. You can use these tools to create feature lines or convert existing polylines, lines, arcs, and 3D polylines into feature lines. You can also create feature lines using corridor, alignments, and stepped offset. The feature line tools can be accessed from the **Feature Line** drop-down in the **Create Design** panel of the **Home** tab. Figure 10-1 shows various tools in the **Feature Line** drop-down.

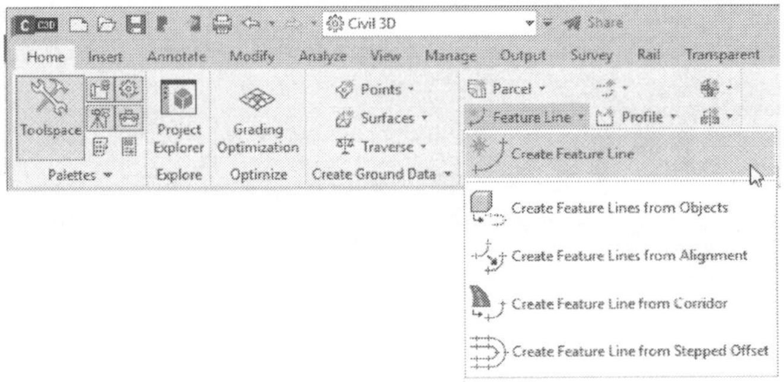

***Figure 10-1** Various tools in the **Feature Line** drop-down*

Creating Feature Lines by Using the Create Feature Line Tool

Ribbon: Home > Create Design > Feature Line drop-down > Create Feature Line
Command: DRAWFEATURELINE

The **Create Feature Line** tool is used to create straight or curved segments as feature lines. To do so, choose the **Create Feature Line** tool from the **Create Design** panel; the **Create Feature Lines** dialog box will be displayed, as shown in Figure 10-2.

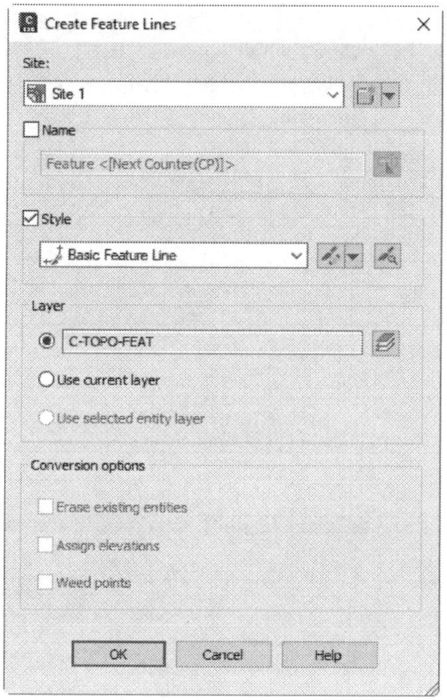

*Figure 10-2 The **Create Feature Lines** dialog box*

Editing Feature Lines

Civil 3D provides you with different tools to help edit the elevation/grade and the geometry of a feature line. The feature lines can also be edited by using the grip editing method.

Editing the Feature Line Geometry

As a feature line is more complex than a line or a polyline feature, Civil 3D provides you with a number of tools to edit the feature line geometry. These tools can be accessed from the **Edit Geometry** panel of the **Modify** tab. This panel consists of tools such as **Break**, **Trim**, **Fillet**, **Smooth**, **Join**, and **Edit Curve**. The **Fillet** and **Smooth** tools in this panel.

Creating Dynamic Feature Lines

In Civil 3D feature line elevations can be obtained from a surface and can also be relative to a surface, so if a surface is updated, the feature line will also be updated. The update behavior of relative feature lines is different depending on whether the feature line was set to be relative to a surface when it was created or it was created at fixed or non-dynamic elevations and then set to be relative.

You can specify that a feature line is relative to a surface while creating or editing it. To do so, choose the **Elevation Editor** tool from the **Edit Elevations** panel of the **Modify** tab; you will be prompted to select the object. Select the required feature line from the drawing; the **PANORAMA** window with the **Grading Elevation Editor** tab chosen will be displayed.

In the **PANORAMA** window, choose the **Elevation from Surface** button to obtain **Set Elevation from Surface** dialog box. Using this command, you can edit the feature lines so that it obtain its elevations from a surface and also relative to that surface. In this dialog box, select the reference surface from the **Select From list** drop-down list. Then, select the **Insert Intermediate Grade Break Points** and **Relative Elevations to Surface** check boxes; the edit box next to **Relative Elevations to Surface** option will become active. Specify the value of relative elevation in the edit box and then choose the **OK** button; **Set Elevation from Surface** dialog box will be closed.

Notice the changes in relative elevation of the feature lines in the **Grading Elevation Editor** tab, as shown in Figure 10-3.

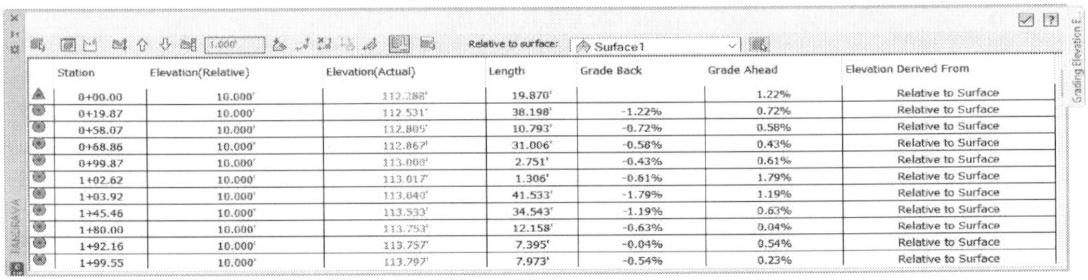

Station	Elevation(Relative)	Elevation(Actual)	Length	Grade Back	Grade Ahead	Elevation Derived From
0+00.00	10.000'	112.388'	19.870'		1.22%	Relative to Surface
0+19.87	10.000'	112.531'	38.198'	-1.22%	0.72%	Relative to Surface
0+58.07	10.000'	112.805'	10.793'	-0.72%	0.58%	Relative to Surface
0+68.86	10.000'	112.867'	31.006'	-0.58%	0.43%	Relative to Surface
0+99.87	10.000'	113.008'	2.751'	-0.43%	0.61%	Relative to Surface
1+02.62	10.000'	113.017'	1.306'	-0.61%	1.79%	Relative to Surface
1+03.92	10.000'	113.040'	41.533'	-1.79%	1.19%	Relative to Surface
1+45.46	10.000'	113.533'	34.543'	-1.19%	0.63%	Relative to Surface
1+80.00	10.000'	113.753'	12.158'	-0.63%	0.04%	Relative to Surface
1+92.16	10.000'	113.757'	7.395'	-0.04%	0.54%	Relative to Surface
1+99.55	10.000'	113.797'	7.973'	-0.54%	0.23%	Relative to Surface

Figure 10-3 Partial view of the changes in relative elevation of the feature line in ***Grading Elevation Editor***

Setting the Properties and Style of the Feature Line

To view and edit the feature line properties and style, select the required feature line and right-click in the drawing; a shortcut menu will be displayed. Choose **Feature Line Properties** from the shortcut menu; the **Feature Line Properties: Site 1** dialog box will be displayed, as shown in Figure 10-4.

Figure 10-4 The ***Feature Line Properties: Site 1*** *dialog box*

Adding Labels to Feature Lines

Ribbon: Annotate > Labels & Tables > Add Labels drop-down > Feature Line > Add Feature Line Labels
Command: ADDFEATURELINELABELS

To add labels to feature lines, choose the **Add Feature Line Labels** tool from the **Labels & Tables** panel; the **Add Labels** dialog box will be displayed. Select the **Line and Curve** option from the **Feature** drop-down list to label lines and curved segments in the feature line. Next, select the **Single Segment** option from the **Label type** drop-down list to label the single segment of the feature line. You can select the **Multiple Segment** option from the **Label type** drop-down list to label all segments of the feature line. Then, select the required option from the **Line label style** and **Curve label style** drop-down list. Choose the **Add** button from the **Add Labels** dialog box; you will be prompted to select a point on the entity. Select the required feature line segment in the drawing; the segment will be labeled.

GRADING OBJECTS

After you have created the feature lines, the next step is to create grading objects from them. The feature line acts as a baseline for these grading objects. Besides using the feature line as a baseline for grading, it can also be added as a breakline in a surface.

The object on which the grading applied is called a grading object. A grading object consists of footprint, daylight line, projection line, and center marker, refer to Figure 10-5. To create a grading object, you need to set the grading criteria which is explained next.

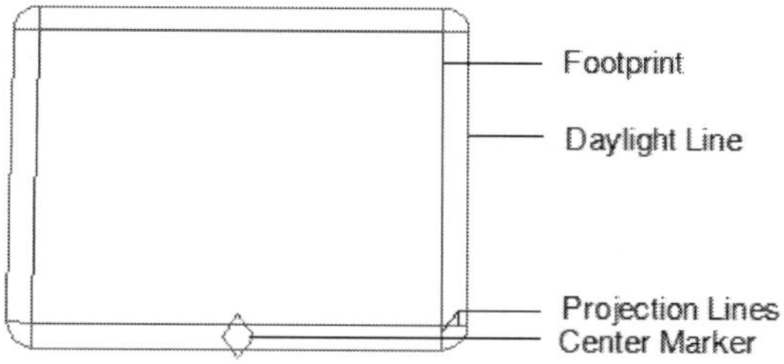

Figure 10-5 *Different parts of a grading object*

Creating Criteria Sets

To create a criteria set, choose the **Settings** tab in the **TOOLSPACE** palette. Next, expand **Grading > Grading Criteria Sets** to view the default criteria set. Right-click on the **Grading Criteria Sets** sub-node and choose the **New** option from the shortcut menu; the **Grading Criteria Set Properties - Grading Criteria Set (1)** dialog box will be displayed. Enter a name of the criteria set in the **Name** edit box. Optionally, you can enter a description in the **Description** text box. Next, choose the **OK** button; the dialog box will be closed and the criteria set will be added to the **Grading Criteria Sets** node. After you have created the criteria set, you need to create and add different criteria to this criteria set. To add the criteria, select the criteria set created and right-click on it; a shortcut menu will be displayed. Choose the **New** option from the shortcut menu displayed; the **Grading Criteria - New Grading Criteria** dialog box will be displayed, as shown in Figure 10-6

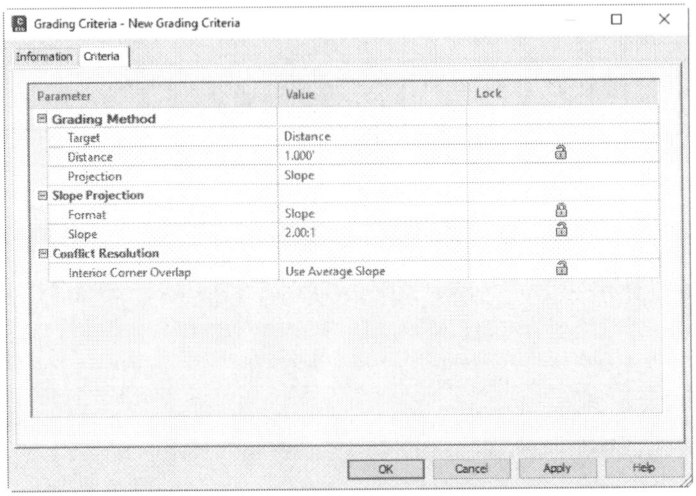

*Figure 10-6 Parameters in the **Criteria** tab of the **Grading Criteria - New Grading Criteria** dialog box*

Creating a Grading Group

You have learned to create feature lines, grading objects, and grading criteria sets. Now, you will learn to create grading groups. Grading groups are created to organize grading into different groups for creating surfaces and earthwork computations.

To create a grading group, choose the **Prospector** tab. Expand the **Sites** node and then expand the required site to which the grading group is to be added. Choose the **Grading Groups** node under the site name and right-click on it; a shortcut menu will be displayed. Choose **New** from the shortcut menu; the **Create Grading Group** dialog box will be displayed, as shown in Figure 10-7.

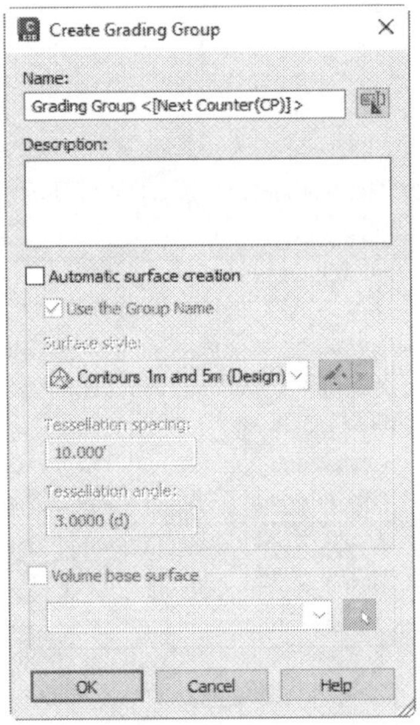

*Figure 10-7 The **Create Grading Group** dialog box*

Using the Grading Creation Tools Toolbar

Ribbon: Home > Create Design > Grading drop-down > Grading Creation Tools
Command: GRADINGTOOLS

The **Grading Creation Tools** toolbar is used to create a new grading, edit the existing grading, and apply the grading criteria and styles. To invoke this toolbar, choose the **Grading Creation Tools** from the **Create Design** panel in the **Home** tab. The **Grading Creation Tools** toolbar contain various tools for grading, as shown in Figure 10-8.

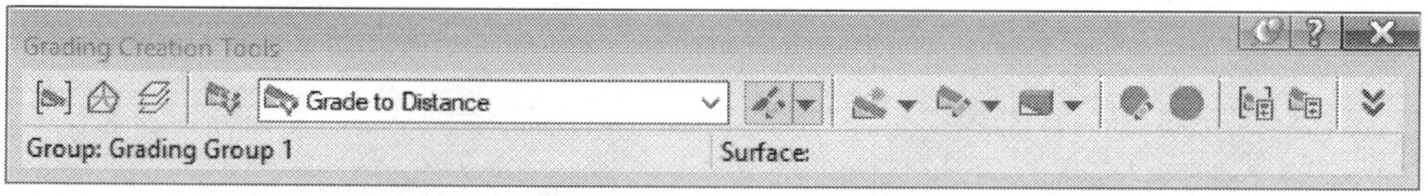

***Figure 10-8** The **Grading Creation Tools** toolbar*

Editing Gradings Using the Grading Editor Tool

Ribbon: Grading > Modify > Grading Editor
Command: GRADINGEDITOR

You can edit a grading by using the **Grading Editor** tab in the **PANORAMA** window. To display the **Grading Editor** tab in the **PANORAMA** window, select grading from the drawing; the **Grading** tab will be displayed in the ribbon. Next, choose the **Grading Editor** tool from the **Modify** panel of this tab; the **Grading Editor** tab in the **PANORAMA** window will be displayed, as shown in Figure 10-9.

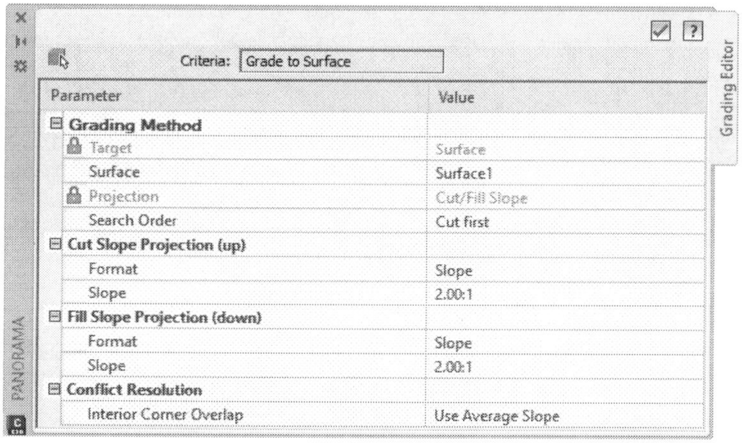

***Figure 10-9** The **Grading Editor** tab in the **PANORAMA** window*

Creating the Grading Infill

Ribbon: Grading > Modify > Creating Grading Infill
Command: CREATEGRADINGINFILL

Grading infill refers to filling of open areas or voids in a grading. The grading infill does not have any criteria. To create a grading infill, select the grading from the drawing; the **Grading** tab will be displayed. Next, choose the **Create Grading Infill** tool from the **Modify** panel; you will be prompted to select the area to infill. Move the cursor to the grading area; the area to be infilled will be highlighted. Click in the required area; the grading infill will be created. Alternatively, invoke the **Create Infill** tool from the **Grading Creation Tools** toolbar to create the grading infill. Figure 10-10 shows the grading without infill and Figure 10-11 shows the grading after creating the grading infill in the empty areas.

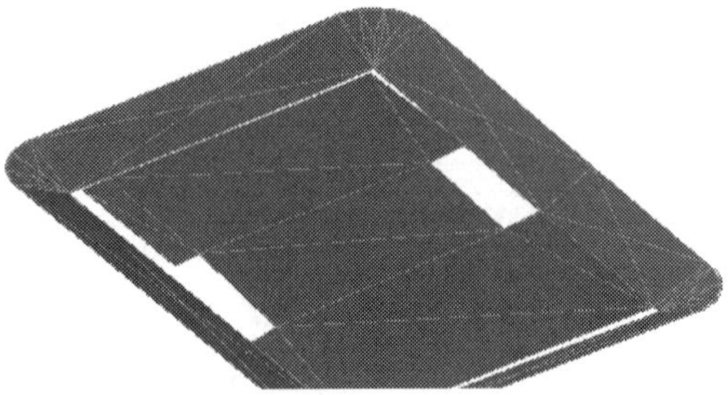

Figure 10-10 *Grading without infill*

Figure 10-11 *Grading after creating the grading infill*

Grading Transition

Civil 3D allows you to add a transition to fill the empty area between gradings. The transition grading is the area that merges with grading on its both sides. You can also add a new grading with a transitional value to the existing grading. To add a transition, invoke the **Create Transition** tool from the **Grading Creation Tool** toolbar, as shown in Figure 10-12; you will be prompted to select a feature line.

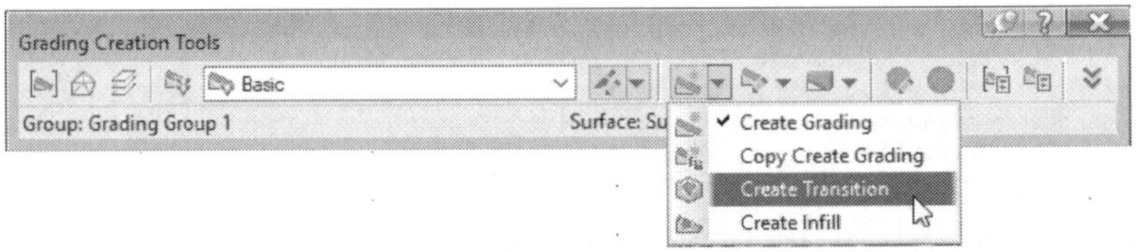

Figure 10-12 *Choosing the* ***Create Transition*** *tool from the* ***Grading Creation Tool*** *toolbar*

Grading Utilities

In AutoCAD Civil 3D, you can use grading utilities to adjust the total cut and fill volumes in a grading design, and to create a detached surface from the grading groups. Before calculating the cut and fill volumes, you need to select the check box in the **Automatic Surface Creation** area of the **Grading Group Properties** dialog box and create a surface.

Grading Style

To create a grading style, choose the **Settings** tab from the **TOOLSPACE** palette and then expand **Grading > Grading Styles** to view the default style. Now, right-click on **Grading Styles** sub-node; a shortcut menu will be displayed. Choose the **New** option from this menu; the **Grading Style - New Grading Style** dialog box will be displayed.

In the **Information** tab of the dialog box, enter a name of the style in the **Name** edit box. Choose the **Center Marker** tab to specify the size of the marker. In this tab, select the required option from the **Size** drop-down list in the **Options** area to calculate the marker size. The **Percentage of Screen** option is the default option selected from the drop-down list. This option allows the size of the marker to be calculated as the fixed percentage of the drawing window. Specify the percentage in the edit box on the right. The **Plotted Size** option is selected from the **Size** drop-down list to specify the size of the marker based on the current drawing unit. On selecting the **Fixed Size** option from the **Size** drop-down list, you can specify the marker size based on the current drawing units. Now, choose the **Slope Patterns** tab from the dialog box, refer to Figure 10-13. This tab is used to select a pattern to represent the grading slopes. Select the **Slope pattern** check box in the **Options** area.

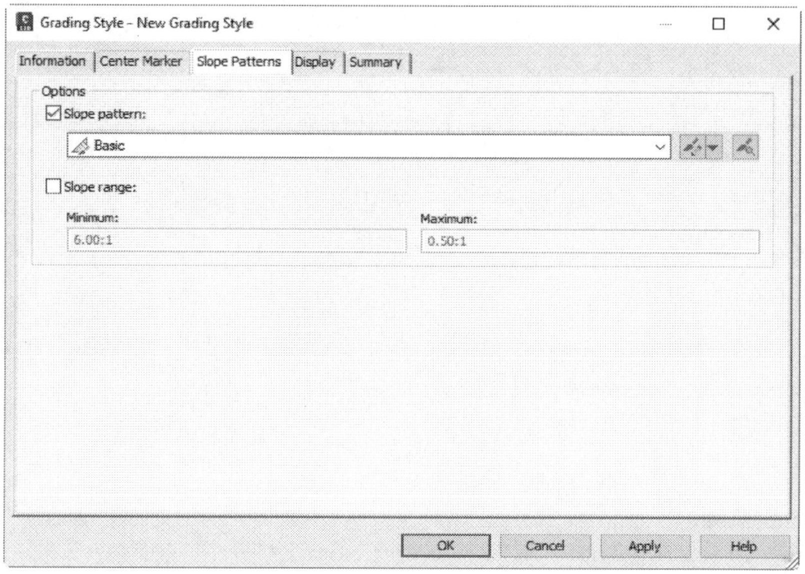

***Figure 10-13** The **Slope Patterns** tab of the **Grading Style - New Grading Style** dialog box*

TUTORIAL

Before starting the tutorial, you need to download and save the tutorial files on your computer. To do so, follow the steps given below:

1. Log on to *www.cadsofttech.com* and browse to *Books > Civil/GIS Books >AutoCAD Civil 3D > AutoCAD Civil 3D 2024 for Novices*. Next, select *c10_c3d_2024_tut.zip* file from the **Tutorial Files** drop-down list. Choose the **Download** button to download the data file.

2. Now, save and extract the downloaded folder to the following location:

   C:\c3d_2024

Note

*While opening the tutorial file, the **PANORAMA** window may appear with an error message. Close this window to proceed further.*

Tutorial 1 Creating Feature Lines

In this tutorial, you then create a feature line style and then feature lines from the existing polylines. Additionally, you will edit feature lines and create grading from them, as shown in Figure 10-14. **(Expected time: 30 min)**

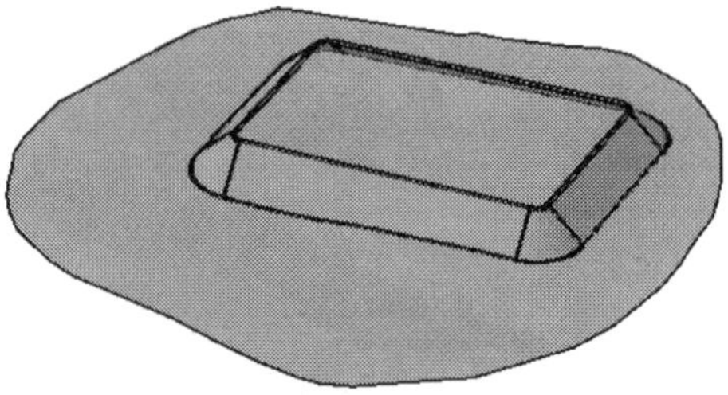

Figure 10-14 *Grading created from the feature lines*

The following steps are required to complete this tutorial:

a. Open the file.
b. Create a new feature line style.
c. Convert polylines into feature lines.
d. Edit the feature lines geometry.
e. Edit the feature lines elevations.
f. Create grading.
g. Save the file.

Opening the File

1. Choose **Open** from the **Application Menu**; the **Select File** dialog box is displayed.

2. In the dialog box, browse to the location *C:\c3d_2024\c10_c3d_2024_tut*.

3. Select the *c10_c3d_2024_tut01.dwg* file and then choose the **Open** button to open the file.

Creating a Feature Line Style

1. Choose the **Settings** tab from the **TOOLSPACE** palette and expand **General > Multipurpose Styles > Feature Line Styles** to view the existing feature line styles.

2. Right-click on **Feature Line Styles**; a shortcut menu is displayed. Choose the **New** option from this menu; the **Feature Line Style - New Feature Line Style** dialog box is displayed. Accept the default name of the style displayed in the **Name** edit box of the **Information** tab.

3. Next, choose the **Profile** tab. In the **Object Display** area of this tab, select the **Basic X** option from the **Beginning Vertex Marker Style** drop-down list, as shown in Figure 10-15.

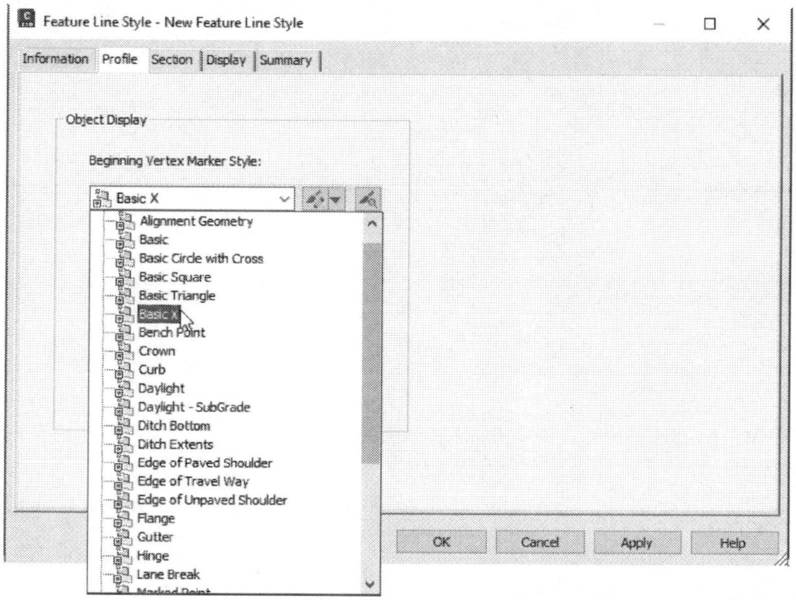

Figure 10-15 *Selecting the **Basic X** option from the **Beginning Vertex Marker Style** drop-down list*

4. Select the **Basic X** option from both the **Internal Vertex Marker Style** and **Ending Vertex Marker Style** drop-down lists.

5. Next, choose the **Section** tab and then select the **Basic X** option from the **Crossing Marker Style** drop-down list.

6. Next, choose the **Display** tab. Ensure that the **Plan** option is selected in the **View Direction** drop-down list.

7. Set the color of the **Feature Line** layer to red and Lineweight to **0.40** mm in the **Component display** list box.

8. Choose the **OK** button; the **Feature Line Style - New Feature Line Style** dialog box is closed and the **New Feature Line Style** style is added to the **Feature Line Styles** node. You will use this line style to display the feature lines.

After creating a feature line style, you need to create feature lines from the polyline objects.

Converting Polylines into Feature Lines

1. Choose the **Create Feature Lines from Objects** tool from **Home > Create Design > Feature Line** drop-down; you are prompted to select objects from the drawing to convert polyline into a feature line.

2. Select the polyline from the drawing, refer to Figure 10-16 and then press ENTER; the **Create Feature Lines** dialog box is displayed.

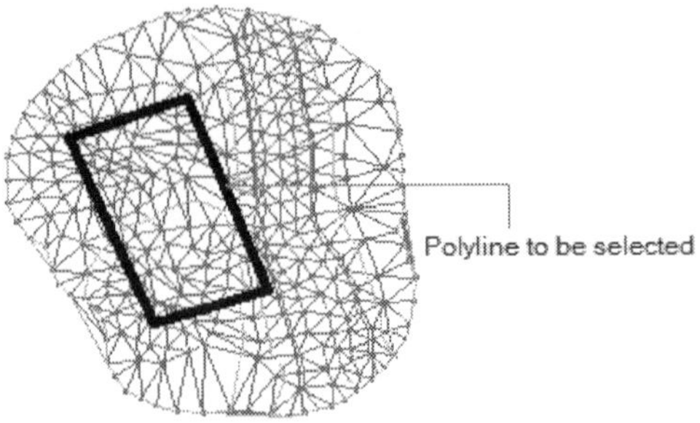

Figure 10-16 *Polyline to be selected from the drawing*

3. Select the **Name** check box in the dialog box; the **Click to edit name template** button is enabled. 

4. Choose the **Click to edit name template** button; the **Name Template** dialog box is displayed.

5. Enter **Building Pad** in the **Name** edit box and choose the **OK** button; the **Name Template** dialog box is closed.

6. Next, in the **Create Feature Lines** dialog box, select the **Style** check box; the drop-down list below the check box is activated.

7. Select the **New Feature Line Style** option from the drop-down list.

8. Choose the **OK** button; the **Create Feature Lines** dialog box is closed and the polylines are converted into feature lines.

Editing the Feature Line Geometry

1. Choose the **Fillet** tool from the **Edit Geometry** panel of the **Modify** tab; you are prompted to select the object. 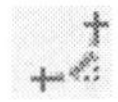

Note

*You can toggle the display of the **Edit Geometry** panel in the **Feature Line** tab by choosing the **Edit Geometry** tool in the **Modify** panel.*

2. Select the building pad feature lines from the drawing; you are prompted to specify a corner or All/Join/Radius.

3. Enter **R** in the command line and press ENTER; you are prompted to specify radius. Enter **15** in the command line and then press ENTER; you are again prompted to specify corner or All/Join/Radius.

4. Enter **A** for all the corners and press ENTER; all the corners of the building pad get filleted, as shown in Figure 10-17. Press ESC to end the command.

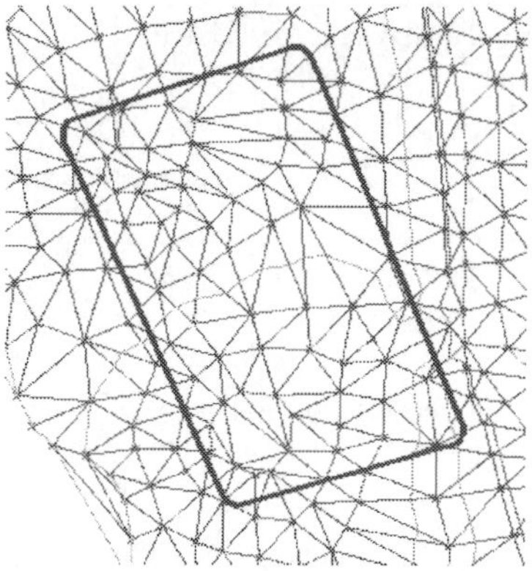

*Figure 10-17 Feature lines of the building pad after using the **Fillet** tool*

Editing the Feature Line Elevation

1. Choose the **Elevation Editor** tool from the **Edit Elevations** panel of the **Modify** tab; you are prompted to select an object from the drawing. Select the **Building Pad** feature line; the **Grading Elevation Editor** tab in the **PANORAMA** window is displayed, as shown in Figure 10-18.

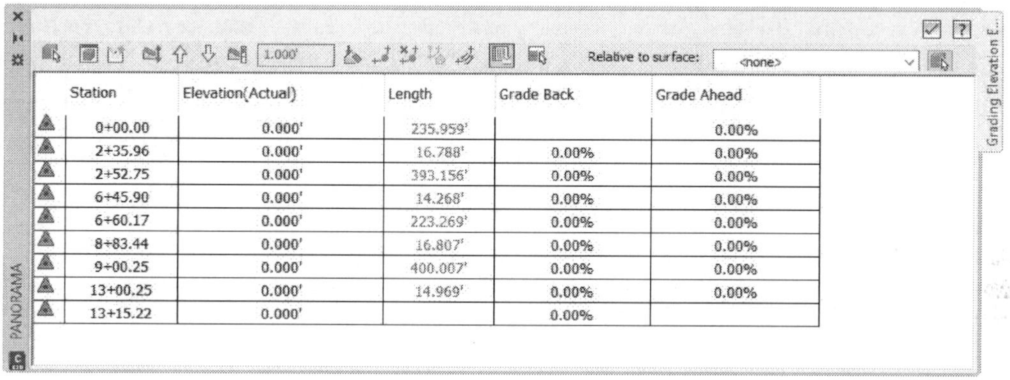

Station	Elevation(Actual)	Length	Grade Back	Grade Ahead
0+00.00	0.000'	235.959'		0.00%
2+35.96	0.000'	16.788'	0.00%	0.00%
2+52.75	0.000'	393.156'	0.00%	0.00%
6+45.90	0.000'	14.268'	0.00%	0.00%
6+60.17	0.000'	223.269'	0.00%	0.00%
8+83.44	0.000'	16.807'	0.00%	0.00%
9+00.25	0.000'	400.007'	0.00%	0.00%
13+00.25	0.000'	14.969'	0.00%	0.00%
13+15.22	0.000'		0.00%	

*Figure 10-18 The **Grading Elevation Editor** tab of the **PANORAMA** window*

2. Choose the **Set Increment** button in the **Grading Elevation Editor** tab of the **PANORAMA** window; the edit box on the right of the button is enabled. 

3. Enter **150.000** in the edit box and then choose the **Raise Incrementally** button from the **Grading Elevation Editor** tab; the elevation of the vertices of all the feature lines is raised by the value specified in the edit box. 

4. Now, choose the green tick mark button; the **PANORAMA** window is closed. 

Creating a Grading

1. Choose the **Grading Creation Tools** tool from **Home > Create Design > Grading** drop-down; the **Grading Creation Tools** toolbar is displayed.

2. Choose the **Set the Grading Group** button from the toolbar; the **Create Grading Group** dialog box is displayed. 

3. Specify **Group 1** as the group name in the **Name** edit box of the dialog box.

4. Choose the **OK** button; the **Create Grading Group** dialog box is closed and the group name is displayed at the bottom of the **Grading Creation Tools** toolbar.

5. In the **Select a Grading Criteria** drop-down list of the **Grading Creation Tools** toolbar, select the **Grade to Surface** option, as shown in Figure 10-19.

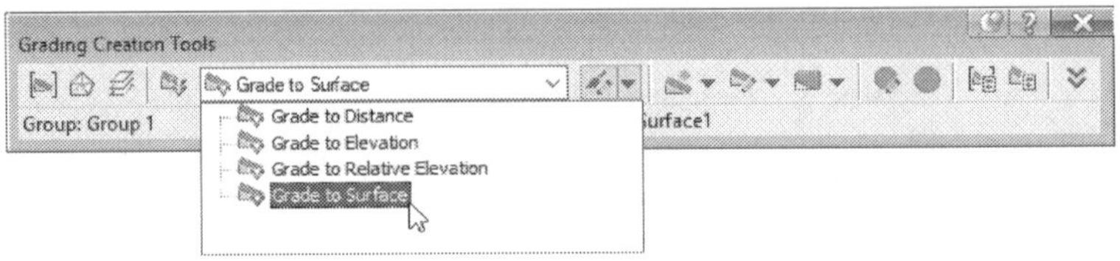

Figure 10-19 *Selecting the **Grade to Surface** option from the **Select a Grading Criteria** drop-down list*

6. Now, choose the **Create Grading** button; you are prompted to select a feature. 

7. Select the feature lines from the drawing; you are prompted to select the grading side. Click in the area outside the closed feature line to specify the grading side; you are prompt whether to apply grading to the entire length or not.

8. Press ENTER to accept the default **Yes** option; you are prompted to specify the cut format.

9. Again, press ENTER; you are prompted to specify the cut slope.

10. Enter **1:1** as cut slope value in the command line and then press ENTER; you are prompted to specify fill format.

11. Again, press ENTER; you are prompted to specify the fill slope. Enter **1:1** as the fill slope value in the command line and then press ENTER; you are prompted to select a feature.

12. Press ESC to exit the command and close the **Grading Creation Tools** toolbar.

13. Choose the **Conceptual** option from **View > Visual Styles > Visual Styles** drop-down.

14. Next, choose the **NE Isometric** option from the **Named Views** panel in the **View** tab to view grading.

15. Use the **ViewCube** tool to view the grading from different angles. Figure 10-20 shows the Conceptual view of the Building Pad grading with NE Isometric orientation.

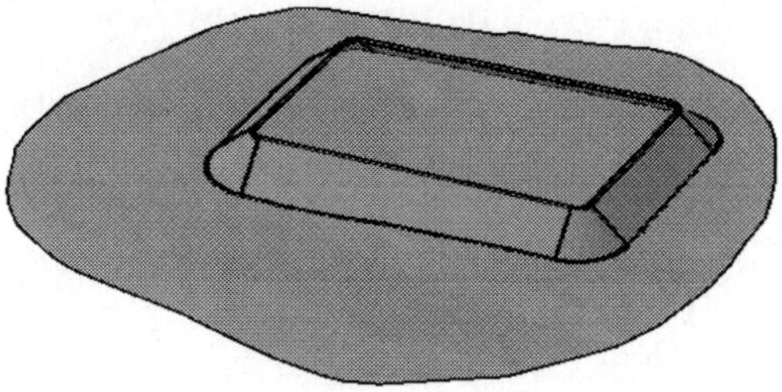

Figure 10-20 The grading in the Conceptual view with the NE Isometric orientation

Saving the File

1. Choose **Save As** from the **Application Menu**; the **Save Drawing As** dialog box is displayed.
2. In this dialog box, browse to the following location:

   C:\c3d_2024\c10_c3d_2024_tut

3. Save the file as *c10_tut01a*.

EXERCISE

1. Log on to *www.cadsofttech.com* and browse to *Books > Civil/GIS Books > AutoCAD Civil 3D > AutoCAD Civil 3D 2024 for Novices*. Next, select *c10_c3d_2024_exer.zip* file from the **Exercise Files** drop-down list. Choose the **Download** button to download the data file.
2. Now, save and extract the downloaded folder to the following location:

   C:\c3d_2024

Note
*While opening the exercise files, the **PANORAMA** window may appear with an error message. Close this window to proceed further.*

Exercise 1

Open the *c10_c3d_2024_ex01.dwg* file and use the baseline in the drawing file. Next, create the grading object using the criteria in the order given below. **(Expected time: 30 min)**

1.	Lane Distance:	**Applied to the main outer circle** **Distance-21**
2.	Lane Elevation:	**Applied to Lane 1** **Elevation-95** **Fill Slope-1:1**
3.	Lane Relative Elevation:	**Applied to Lane 2** **Relative Elevation-7**

4. Lane Surface: **Applied to Lane 3**
Surface
Fill Slope-2:1
Cut Slope-2:1

Apart from the values given, accept all the default values of the properties when you are prompted to do so. Save the file as *c10_ex01a*. The final grading for this exercise is shown in Figure 10-21.

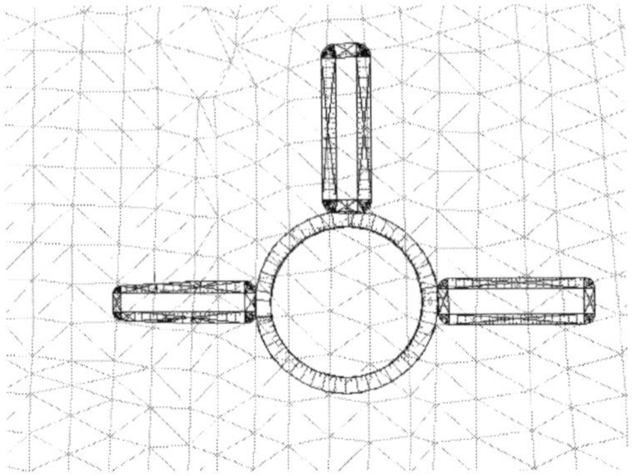

Figure 10-21 *The lanes and the main circle after grading*

Chapter 11

Pipe Networks

Learning Objectives

After completing this chapter, you will be able to:

- *Understand the pipe network object and its parts*
- *Understand the part builder and the part catalog*
- *Create the network parts list*
- *Understand part rules*
- *Create the pipe network using the network layout tools*
- *Create the pipe network from objects*
- *Draw pipe network parts in a profile view*
- *Edit pipe networks*
- *Add labels to the pipe network*

OVERVIEW

Pipe networks are an integral part of a design site. A well-designed pipe network, be it a sewage system, storm drainage system, or a water supply system, is an essential aspect of a designed land site. For example, a storm drainage system avoids flooding in an area which occurs due to excessive surface runoff. Runoff is part of the rainwater, snow melt, or irrigation water that occurs when there is more water than land can absorb, it flows over the ground and returns back to the streams or rivers. The calculation of runoff is an important factor in the site design as the runoff causes erosion of soil, leading to a change in the landscape. Also, excessive runoff may lead to the flooding of an area. The engineers should be aware of the retention capacity of the site and the amount of runoff that the site will detain after being developed. If the excess of runoff is not checked, it may lead to floods and other disturbances.

Hydraulic systems, storm sewers, sanitary sewers, detention ponds and culverts are some of the features that are designed to deal with the flow of water and help diverting the excess runoff to avoid flood. AutoCAD Civil 3D has the ability to analyze and design such hydraulic systems and help the engineers enhance and improve their designs. In AutoCAD Civil 3D, the pipe network objects allow you to draft 2D and 3D models of the hydraulic systems along with the visual effects and useful data. AutoCAD Civil 3D enables you to perform various water management and analysis tasks using three application extensions that are installed on your computer with the installation of Civil 3D. These application extensions are briefly discussed next:

Hydraflow Express Extension: This application allows you to perform hydraulic or hydrology studies on culverts, inlets, weirs, and channels. You can model and design culverts with different shapes using this application extension.

Hydraflow Hydrographs Extension: This extended application of AutoCAD Civil 3D allows you to design and model the detention ponds and watersheds.

Hydraflow Storm Sewers Extension: This application of Civil 3D allows you to import the pipe networks from AutoCAD Civil 3D and modify the pipe designs, perform designs calculations, modify slopes, and then export the modified pipe network back to Civil 3D.

These extensions can be accessed and opened by choosing the **Launch Storm Sewers / Launch Hydrographs / Launch Express** tool from the **Design** panel of the **Analyze** tab. In this chapter, you will learn about pipe networks and other related topics. First, you must get familiar with different terms related to the pipe network object that are discussed next.

PIPE NETWORK

The pipe network object is a collection of different pipes and structures, such as manholes and catch basins that form a pipe system. The pipe network also specifies and maintains the relationship between different parts of the network and other Civil 3D objects, such as surfaces, alignments, and labels. The pipe networks are added in plan views and can consist of both pipes and structures or pipes and structures individually.

AutoCAD Civil 3D provides you with various inbuilt styles to display the pipes and structures with visual effects and data. You can use these styles to display the pipes and structures in plan, profile, or section views as required.

The pipe network objects are listed in the **Networks** sub-node of the **Pipe Networks** node in the **Prospector** tab of the **TOOLSPACE** palette. The information about the pipe network is displayed in the **Prospector Item View** of the **Prospector** tab. Figures 11-1 and 11-2 show a 2D view and a conceptual 3D view of a pipe network, respectively. The pipe network consists of various components and parameters such as pipes, structure and null structure. These components and parameters are discussed next.

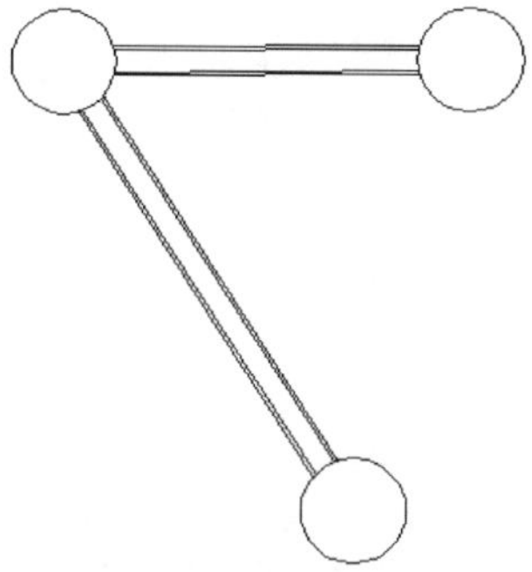

Figure 11-1 *The 2D view of a pipe network*

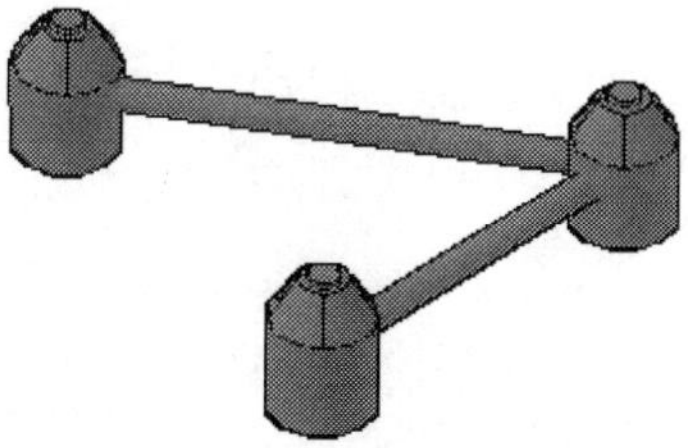

Figure 11-2 *The 3D view of a pipe network*

Pipes

A pipe object can be defined as a tube or a conduit that carries fluid under the gravitational force or pressure applied externally (pumped fluids). Pipes are available in various sizes and shapes such as rectangular, elliptical, oval, or circular. A pipe can be made from different materials such as steel, concrete, PVC, or iron. To create a pipe network in Civil 3D, you can use the default pipes provided in the Civil 3D library or create your own pipes as required. Civil 3D allows you to create and use different pipe styles to control the appearance of pipes in a plan, section, or profile view. The pipe styles are created and managed in the **Settings** tab of the **TOOLSPACE** palette.

Structures

A structure represents the manholes, catch basins, headwalls, and other parts of a pipe network. The structures are more complex than pipes. AutoCAD Civil 3D provides some in-built structures with specified dimensions and shapes. There are four types of structures available in Civil 3D: **Inlets-Outlets**, **Junction Structures with Frames**, **Junction Structures without Frames**, and **Null** structures. You can modify these structures as required and then use them in the network. Like pipes, structures too have different styles that control their display in the drawing.

Null Structures

The null structure is a special type of structure created at the junction of two pipes. This type of structure is automatically added in the pipe network when a pipe is connected with the other pipe without using any other type of structure. The main use of the null structure is to connect the pipes, as shown in Figure 11-3. You can hide the null structures in a drawing view by creating and assigning different styles to them.

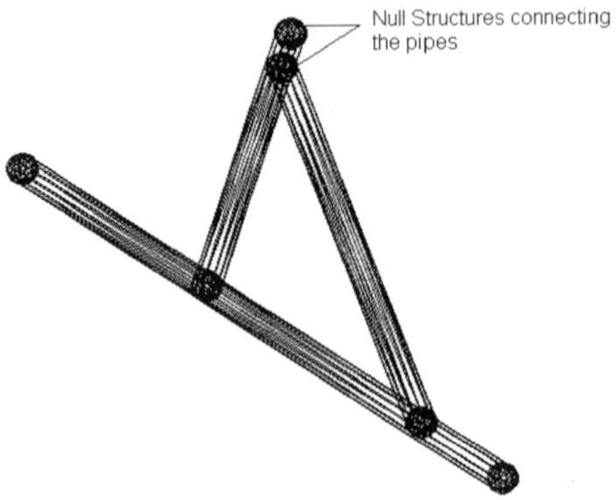

Figure 11-3 *The null structures connecting the pipes*

Note

The names of the pipes, structures, and the null structure objects cannot be added to the ***Prospector*** *tab of the* ***TOOLSPACE*** *palette, but the information related to them can be viewed in the list view. To do so, select a pipe network in the* ***Prospector*** *tab; the information about this network will be displayed in the list view, as shown in Figure 11-4.*

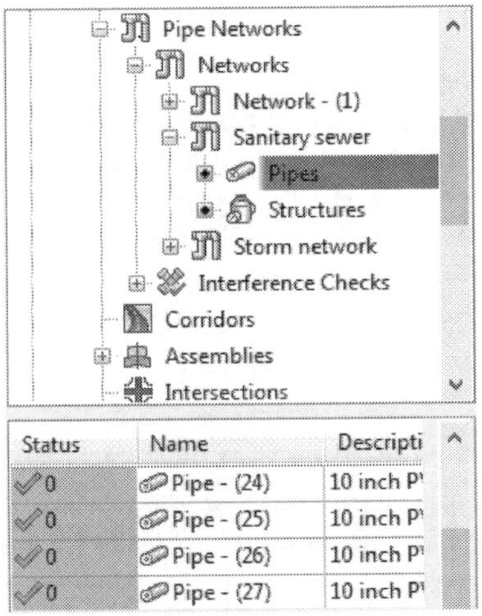

Figure 11-4 *The* ***Prospector*** *list view displaying information about* ***Sanitary sewer*** *pipe network*

Part Catalog

AutoCAD Civil 3D provides you with the part catalog consisting of various in-built parts (pipes and structures) of different sizes, shapes, materials, and behavior used in a pipe network. After deciding the parts that you want to use to create a design, you can find those parts in the part catalog. The part catalog is automatically saved at the location *C:\ProgramData\Autodesk\C3D 2024\enu\Pipes Catalog* while installing AutoCAD Civil 3D.

A Civil 3D part catalog has two basic domains to select **Pipes** and **Structures**. Both the **Pipes** and the **Structures** domains consist of Metric and Imperial pipe catalogs in the corresponding folders. These folders together form the content of the part catalog. To display and use this catalog, browse to *C:\ProgramData\Autodesk\C3D 2024\enu\Pipes Catalog\US Imperial Pipes*. Each sub-folder in this folder can be expanded to view the type of pipes. Select the required pipe from the left pane of the catalog to view the image, specifications, and available sizes of the selected pipe type.

Similarly, open the *US Imperial Structures.html* file from the **US Imperial Structures** folder. The default structure catalog consists of four different categories representing four types of structures.

Parts List

A parts list is a subset of components which are derived from a master catalog of components of a pipe network. A parts list is a collection of required parts (pipes and structures) of a pipe network. The parts list enables you to store the selected pipes, structures, styles, render materials, rule sets, and so on separately in a parts list so that you do not need to navigate and look for the required parts in the part catalog. Thus, the parts list saves time and helps you to use the required parts to create the pipe network in an organized way. You can create different parts list for different projects based on the project requirement. Civil 3D has some default parts list that are stored in the Civil 3D template. However, you can also create your own multiple parts list and then use them in the project.

The parts lists are added in the **Part Lists** sub-node of the **Pipe Network** node in the **Settings** tab of the **TOOLSPACE** palette. Before creating a parts list, you need to specify the part catalog from which the parts list would be derived. Specify the catalog using the **Set Pipe Network Catalog** tool as described earlier.

Part Builder

AutoCAD Civil 3D allows you to create and modify pipe network parts available in the part catalog of Civil 3D using the **Part Builder** tool. You can also build the customized network parts using the Part Builder. Remember that the Part Builder should be used by the experienced Civil 3D users who are well acquainted with the parametric modeling and pipe network parts.

Note

Before you start using the Civil 3D Part Builder, save a copy of the part catalog to retain the default parts. As mentioned earlier, the catalogs can be accessed from C:\ProgramData\Autodesk\C3D 2024\enu\Pipes Catalog\US Imperial Structures. This will help you to avoid deleting or changing the default parts in the catalogs.

Part Rules

The part rules are a set of rules that govern the elevation, location, and the way the pipes are connected to the structures. These rules also help you to keep a check and warn you if certain criteria are violated after creating or editing the pipe network parts. Civil 3D has some built in set of rules that are automatically applied to each part of the pipe network. You can also create your own set of rules and use them to design the pipe network.

The part rules are divided into two parts: Pipe Rules and Structure Rules. The combination of these rules is called a rule set. The rules can be applied to the pipe network parts as a rule set.

CREATING A PIPE NETWORK

AutoCAD Civil 3D provides a set of tools to design pipe network for transporting fluid using earth's gravitational force. The objects in the network can be created by using the pipe network layout tools or by using the objects such as polyline, line, and feature line. The methods of creating the pipe network are discussed next.

Using the Network Layout Tools

Ribbon:	Home > Create Design > Pipe Network drop-down > Pipe Network Creation Tools
Command:	CREATENETWORK

To create a pipe network using the pipe network layout tools, choose the **Pipe Network Creation Tools** tool from the **Create Design** panel; the **Create Pipe Network** dialog box will be displayed, as shown in Figure 11-5. This dialog box is used to specify the information about the pipe network to be created.

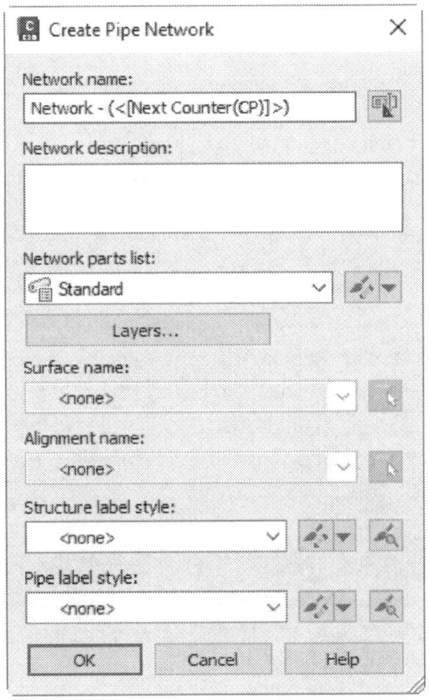

*Figure 11-5 The **Create Pipe Network** dialog box*

Pipes and Structures Tool

This tool allows you to create a pipe network with both the pipes and structures parts. To create a pipe network, choose the **Pipes and Structures** tool from the **Network Layout Tools** toolbar and create a pipe network with a series of structures connected to the pipes. Figure 11-6 shows a pipe network with both pipes and structure parts.

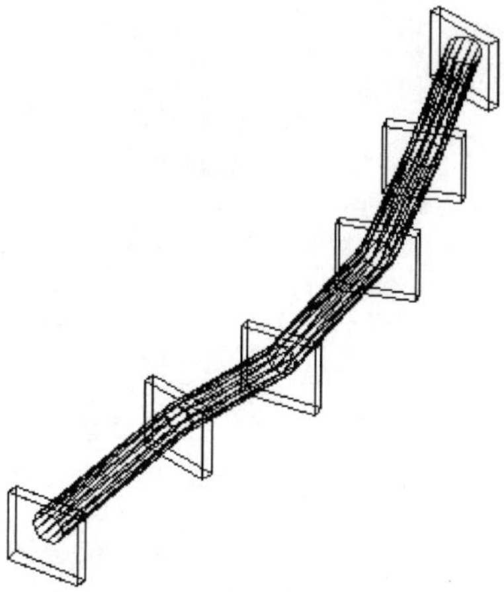

Figure 11-6 *A pipe network with both pipes and structure parts*

Pipes Only Tool

The **Pipes Only** tool allows you to create a network by only drawing the pipes. In this case, the network will have only the null structures joining the pipes. To create a pipe network by using only pipes, choose the down-arrow on the right of the **Pipes and Structures** tool to display a flyout. Next, and choose the **Pipes Only** tool from the flyout displayed. Next, click in the drawing to specify the first pipe point. Follow the prompts and click in the drawing again to specify other pipe points. Figure 11-7 shows a part of the network with only the pipes connected with the null structures.

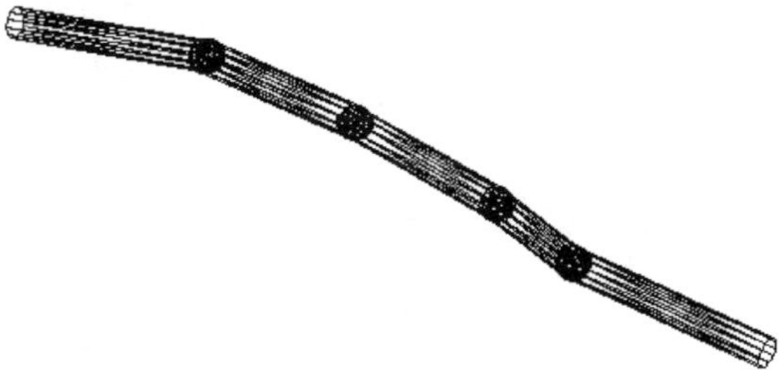

Figure 11-7 *A part of the network with only the pipes connected with the null structures*

Structures Only Tool

This tool allows you to draw only the structures in an existing pipe network. To use this mode, choose the down-arrow next to the **Pipes and Structures** tool from the **Network Layout Tools - <network name>** toolbar and then choose the **Structures Only** tool from the flyout displayed. Next, click in the drawing at the required location to specify the structure insertion points for the structures. Figure 11-8 shows the structures added to the existing pipe.

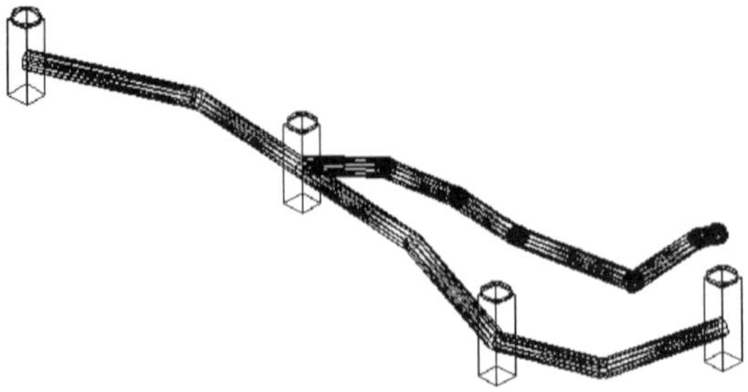

Figure 11-8 *The structures added to the pipe*

After you have created the pipe network using any of the three tools, you can check if any of the part rule is violated. To do so, choose the **Pipe Network Vistas** button from the toolbar; the **PANORAMA** window with the **Pipes** and **Structures** tabs will be displayed. To check the status of the pipe rule, choose the **Pipes** tab in the **PANORAMA** window. In case the rules have been violated, the violation icon will be displayed in the **Status** column for the pipes which have violated the rules, as shown in Figure 11-9.

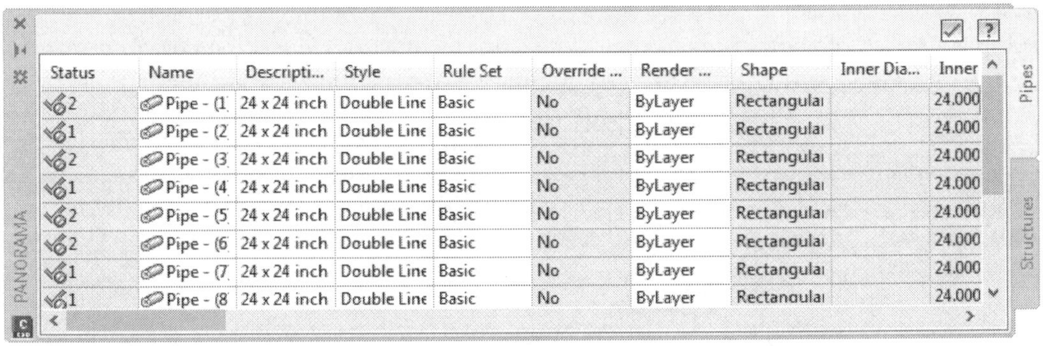

Status	Name	Descripti...	Style	Rule Set	Override ...	Render ...	Shape	Inner Dia...	Inner
2	Pipe - (1	24 x 24 inch	Double Line	Basic	No	ByLayer	Rectangular		24.000
1	Pipe - (2	24 x 24 inch	Double Line	Basic	No	ByLayer	Rectangular		24.000
2	Pipe - (3	24 x 24 inch	Double Line	Basic	No	ByLayer	Rectangular		24.000
1	Pipe - (4	24 x 24 inch	Double Line	Basic	No	ByLayer	Rectangular		24.000
2	Pipe - (5	24 x 24 inch	Double Line	Basic	No	ByLayer	Rectangular		24.000
2	Pipe - (6	24 x 24 inch	Double Line	Basic	No	ByLayer	Rectangular		24.000
1	Pipe - (7	24 x 24 inch	Double Line	Basic	No	ByLayer	Rectangular		24.000
1	Pipe - (8	24 x 24 inch	Double Line	Basic	No	ByLayer	Rectangular		24.000

Figure 11-9 *The violation icon displayed in the* ***Status*** *column for the pipes violating the pipe rules*

Similarly, choose the **Structures** tab in the **PANORAMA** window and check the status of the structures in the **Status** column. You can also use the **PANORAMA** window to edit the pipe and structure parameters such as name, description, render material, and style.

Alternatively, you can also check the status of the pipes in the **Prospector** list view. To do so, expand **Pipe Networks > Networks** in the **Prospector** tab and then expand the required network. Next, select **Pipes** or **Structures** from the expanded node to view the respective information in the list view. You can use the **Network Layout Tools - <network name>** to draw the pipe network according to the design and the project requirement. This method provides you the flexibility of selecting the network parts, part styles, and the mode of creating the pipe network.

Creating Pipe Networks from Objects

Ribbon: Home > Create Design > Pipe Network drop-down > Create Pipe Network from Object
Command: CREATENETWORKFROMOBJECT

You can create a pipe network from the existing AutoCAD entities such as line, polylines, splines, arcs, and feature lines. AutoCAD Civil 3D converts these entities to a pipe network and then inserts structures at the end points or vertices of these drawing entities. To create a pipe network from an object, choose the **Create Pipe Network from Object** tool from the **Create Design** panel; you will be prompted to select an object or an Xref in the command line. Select the existing entity in the drawing; you will notice an arrow displayed on the selected entity showing the direction of flow. Press ENTER to accept the default direction or enter **R** in the command line and then press ENTER; the direction of flow will be reversed. Again, press ENTER to display the **Create Pipe Network from Object** dialog box, as shown in Figure 11-10.

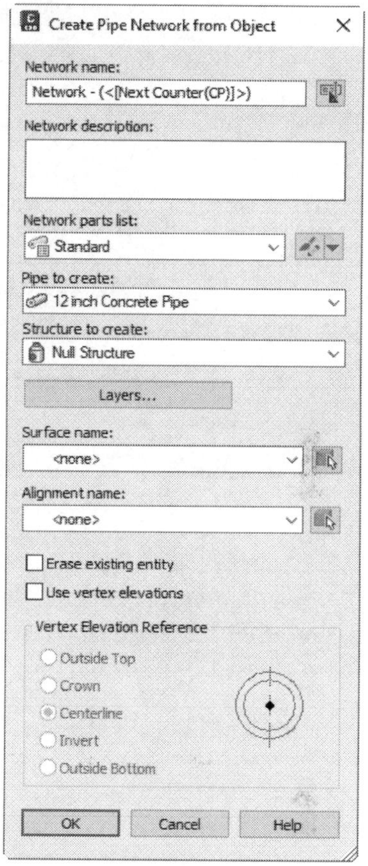

***Figure 11-10** The **Create Pipe Network from Object** dialog box*

The options in the dialog box are the same as those explained in the **Create Pipe Network** dialog box. You can select the **Erase existing entity** check box to remove the entity from which the network is created. Select the **Use vertex elevations** check box to use the elevations of vertices along the selected entity to set the elevations of the pipes created in the network. On selecting this check box, the options in the **Vertex Elevation Reference** area will become active. You can select an option from this area to specify the location on the pipe where the vertex elevation is located. Note that when you select an option from this area, the rules set for the pipe network will no longer be applicable.

After you have specified the required information in the **Create Pipe Network from Object** dialog box, choose the **OK** button; the dialog box will be closed and a pipe network will be created with the structures inserted at the vertices of the selected entity.

DRAWING PIPE NETWORK PARTS IN A PROFILE VIEW

You can view the parts of an existing pipe network in a profile view to evaluate the vertical positions of the network parts based on your design. This will help you to visualize and compare the pipe network elevation with those of the road design. You can view the entire pipe network or the required pipe network parts in a profile or a section view.

To draw the network parts in a profile view, you need to have a pipe network, a profile view, and an alignment providing the station data in your drawing. To draw, select the profile view; the **Profile View: <profile view name>** tab will be displayed. Next, choose the **Draw Parts in**

Profile View tool from the **Launch Pad** panel of the **Profile View :<profile view name>** tab; you will be prompted to select the pipe network or a part of pipe network that you want to display in the profile view. Select the required network part in the drawing and press ENTER; you are now prompted to select the profile view in which you want to display the network parts. Select the required profile view in which you want to display the network parts; the selected network parts will be displayed in the profile view, as shown in Figure 11-11.

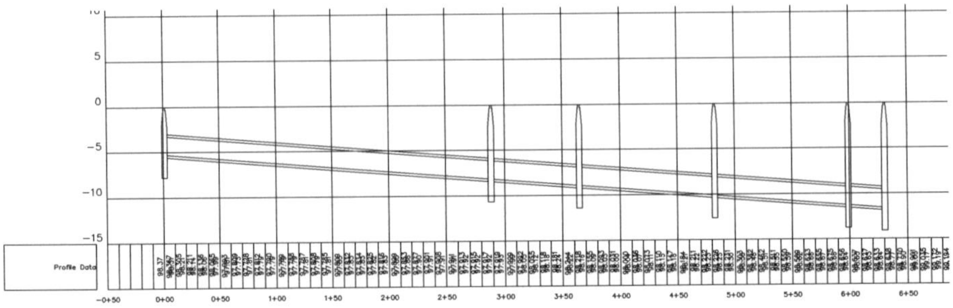

Figure 11-11 *The network parts displayed in the profile view*

NETWORK PROPERTIES

To access and edit the pipe network properties such as network name, labels, layers, profile and section view settings, select any part of the network in the drawing and right-click; a shortcut menu will be displayed. Choose the **Network Properties** option from the shortcut menu displayed; the **Pipe Network Properties - <network name>** dialog box will be displayed, refer to Figure 11-12. The **Information** tab of this dialog box displays the name of the selected pipe network. Modify the name in the **Name** edit box, if required.

Choose the **Layout Settings** tab to view and edit the parameters used for the pipe network in the plan view. The **Labels** area displays the label styles used to label the pipes and structures in the plan view, refer to Figure 11-12. Select the required options from the **Structure plan label style** and **Pipe plan label style** drop-down lists. Select an option from the **Network parts list** drop-down list to define the part list used for the selected pipe network. You can select the required part list from the drop-down list or create a new part list. To create a new part list, choose the down arrow from the button on the right of this drop-down list; a flyout will be displayed. Select the **Create New** option from the flyout to create a new part list. The **Default object reference** area displays the name of the surface and alignment objects to which the pipe network is referenced. If the network is not referenced to any surface or alignment, the **none** option will be selected by default in the **Surface name** and **Alignment name** drop-down lists in this area.

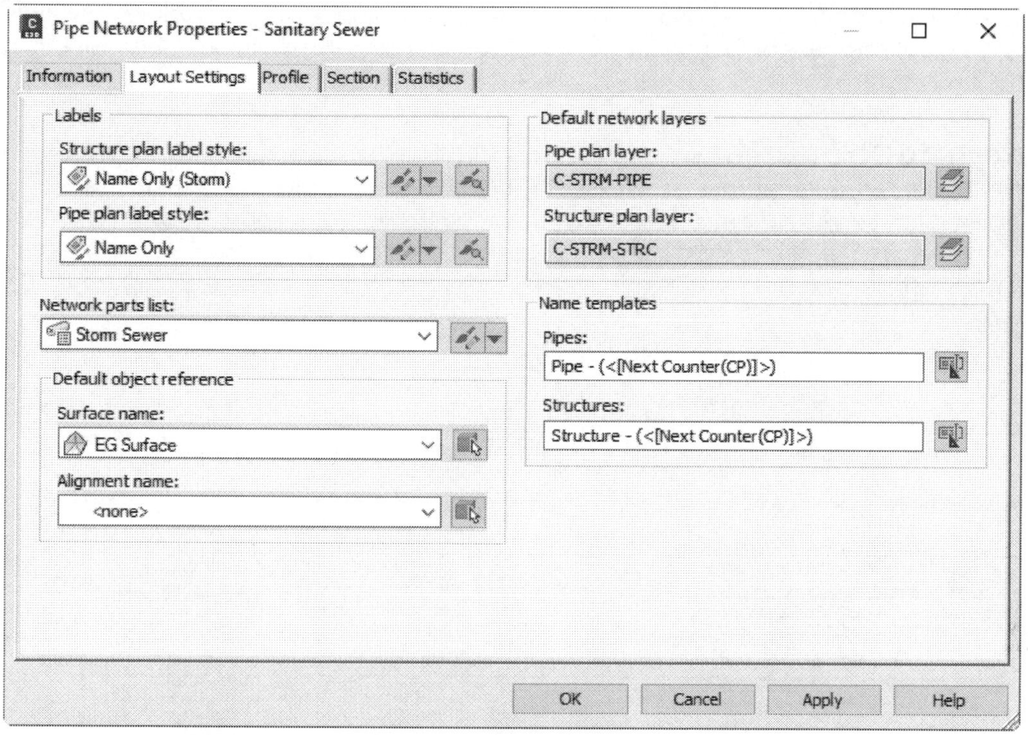

Figure 11-12 *The* ***Pipe Network Properties*** *dialog box*

The **Default network layers** area displays the layers in which the pipes and structures will be drawn in the plan view. Choose the **Select Layers** buttons on the right in this area to display the **Object Layer** dialog box. You can use this dialog box to change the default layer. The **Name templates** area displays the default names of the templates used for creating the pipes and structures in the network. Choose the button on the right of the **Pipes** or **Structures** edit box to display the **Name Template** dialog box. Enter a name for the pipe or structure in the **Name** edit box in this dialog box and choose the **OK** button to exit.

The **Profile** tab displays the label styles used to label the pipes and structures in the network in the profile view. If no styles are used, the **none** option will be selected by default in the **Structure profile label style** and the **Pipe profile label style** drop-down lists in the **Labels** area. You can select the required option from these drop-down lists to specify the label style. The **Default profile layers** area displays name of the default layers on which the pipes and structures will be drawn in the profile view. Choose the button on the right of the **Pipe profile layer** text box; the **Object Layer** dialog box will be displayed. You can modify the default layers by using the options in the dialog box.

The **Section** tab displays the name of the default layer to display the network objects in a section view. You can modify the default layer by choosing the **Select Layers** button next to the **Pipe network section layer** text box in the **Default section layers** area.

The **Statistics** tab of the **Pipe Network Properties - <network name>** dialog box displays the network statistics for the selected network. Expand the **General**, **Pipes**, and **Structures** categories in this tab to view their statistics. After you have viewed or edited the network properties, choose the **OK** button to exit the **Pipe Network Properties - <network name>** dialog box.

Structure Properties

To view and edit the properties of a structure in a network, select the required structure from the drawing and right-click; a shortcut menu will be displayed. Choose **Structure Properties** from the shortcut menu; the **Structure Properties - <structure name>** dialog box will be displayed, refer to Figure 11-13.

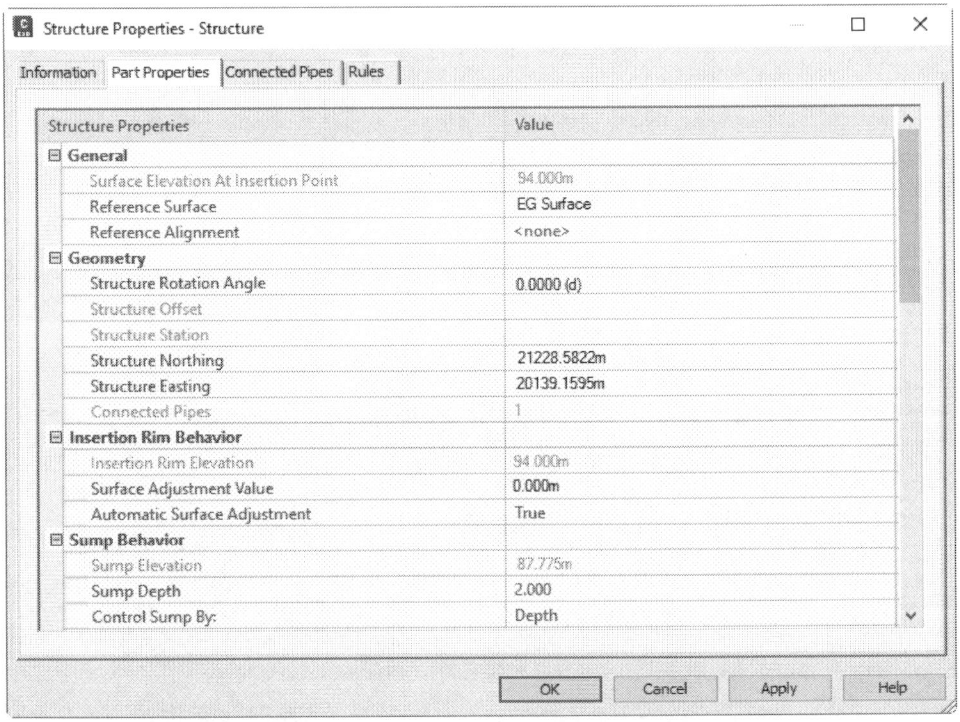

*Figure 11-13 The **Structure Properties** dialog box*

Pipe Properties

To view and edit the pipe properties, select the required pipe part from the drawing and right-click; a shortcut menu will be displayed. Next, choose the **Pipe Properties** option from the shortcut menu; the **Pipe Properties - <pipe name>** dialog box will be displayed.

EDITING A PIPE NETWORK IN PLAN LAYOUT

Civil 3D provides you with different methods for editing the pipe networks in a plan layout. You can use the **Network Layout Tools** toolbar, **TOOLSPACE** Item View, and the grip editing methods to edit the pipe network as required. These editing methods are discussed next.

Using the Pipe Network Layout Tools Toolbar

Ribbon: Pipe Networks: <Network Name> > Modify > Edit Pipe Network
Command: EDITNETWORK

To edit the pipe network by using the **Pipe Network Layout tools** toolbar, choose the **Edit Pipe Network** tool from the **Modify** panel of the **Pipe Networks: <Network Name>** tab; the **Network Layout Tools - <network name>** toolbar will be displayed. You can choose various tools from this toolbar to add, change, or delete the network parts. You can also modify the associated surface, alignment, and the part list using this toolbar.

ADDING PIPE NETWORK LABELS

Ribbon: Annotate > Labels & Tables > Add Labels drop-down > Pipe Network > Add Pipe Networks Labels
Command: ADDNETWORKLABELS

You can label the pipe network or some parts of the network using the in-built label styles. The label styles can be created and managed in the **Settings** tab of the **TOOLSPACE** palette. You can access the pipe and structure labels styles in the **Pipe** and **Structure** nodes respectively. The labels can be added while creating the network or they can be added after the pipe network is created. To add the label while creating the network, select the required label style in the **Create Pipe Network** dialog box.

To add the pipe network labels after the pipe network is created, choose the **Pipe Network** tool from the **Add Labels** drop-down in the **Labels & Tables** panel of the **Annotate** tab; a flyout will be displayed with different tools to label the pipe network. Choose the **Add Pipe Network Labels** tool from the flyout displayed, as shown in Figure 11-14. On doing so, the **Add Labels** dialog box will be displayed.

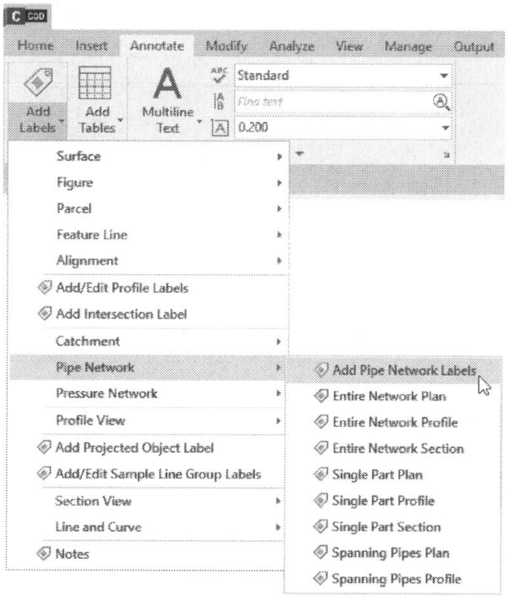

Figure 11-14 *Choosing the* ***Add Pipe Network Labels*** *tool from the flyout*

ADDING TABLES

You can view the network pipe and structure data in a tabular form and add the tables to the drawing. These tables help you to view the consolidated information about the structure and pipe objects that are used to draw the pipe network.

Adding a Structure Table

Ribbon:	Annotate > Labels & Tables > Add Tables drop-down > Pipe Network > Add Structure
Command:	ADDNETWORKSTRUCTTABLE

To add the table for structures, choose the **Pipe Network** option from the **Add Tables** drop-down; a flyout will be displayed. Choose the **Add Structure** option from the flyout displayed; the **Structure Table Creation** dialog box will be displayed, as shown in Figure 11-15. In this dialog box, select the required option from the **Table style** drop-down list. You can also create a new style for structure table. To do so, choose the down arrow on the right of the **Table style** drop-down list; a flyout will be displayed. Choose **Create New** option from the flyout; the **Table Style - New Structure Table Style** dialog box will be displayed. Specify the required options in this dialog box and choose the **OK** button to close the dialog box.

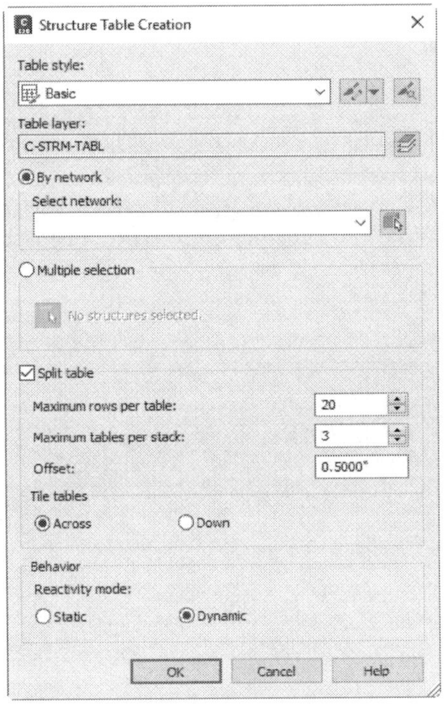

Figure 11-15 *The* ***Structure Table Creation*** *dialog box*

Adding a Pipe Table

Ribbon: Annotate > Labels & Tables > Add Tables drop-down > Pipe Network > Add Pipe
Command: ADDNETWORKPIPETABLE

To add the tables for the pipe, choose the **Pipe Network** option from the **Add Tables** drop-down; a flyout will be displayed with different options. Choose the **Add Pipe** option from the flyout displayed; the **Pipe Table Creation** dialog box will be displayed. The options in this table are same as discussed for the **Structure Table Creation** dialog box. After you have specified the settings and selected the required network or the pipe objects, choose the **OK** button; the **Pipe Table Creation** dialog box will be closed and you will be prompted to specify the upper left corner of the table. Click in the drawing to specify the table corner; the pipe table will be added to the drawing displaying the pipe details, as shown in Figure 11-16. Like the Structure table, you can configure the settings for the pipe table in the table style and display the required information about the pipe object in the table.

Pipe Table			
Pipe Name	Size	Length	Slope
Pipe - (10)	24.000	287.117	1.00%
Pipe - (11)	24.000	76.153	1.00%
Pipe - (12)	24.000	117.107	1.00%
Pipe - (13)	24.000	116.654	1.00%
Pipe - (14)	54.000	31.906	1.00%

Figure 11-16 *The Pipe table displaying the pipe details*

TUTORIAL

Before starting the tutorial, you need to download and save the tutorial files on your computer. To do so, follow the steps given below:

1. Log on to *www.cadsofttech.com* and browse to *Books > Civil/GIS Books >AutoCAD Civil 3D > AutoCAD Civil 3D 2024 for Novices*. Next, select *c07_c3d_2024_tut.zip* file from the **Tutorial Files** drop-down list. Choose the **Download** button to download the data file.

2. Now, save and extract the downloaded folder to the following location:

   C:\c3d_2024

Note

*While opening the tutorial file, the **PANORAMA** window may appear with an error message. Close this window to proceed further.*

Tutorial 1 Sanitary Sewer Network

In this tutorial, you will create a sanitary sewer network, as shown in Figure 11-17, using the **Network Layout Tools** toolbar. You will also create a parts list and part label styles.

(Expected time: 30 min)

The following steps are required to complete this tutorial:

a. Open the file.
b. Create the parts list.
c. Create the structure label style.
d. Add another component to the label.
e. Create the sanitary sewer network using the **Network Layout Tools** toolbar.
f. Reference the label text.
g. Save the file.

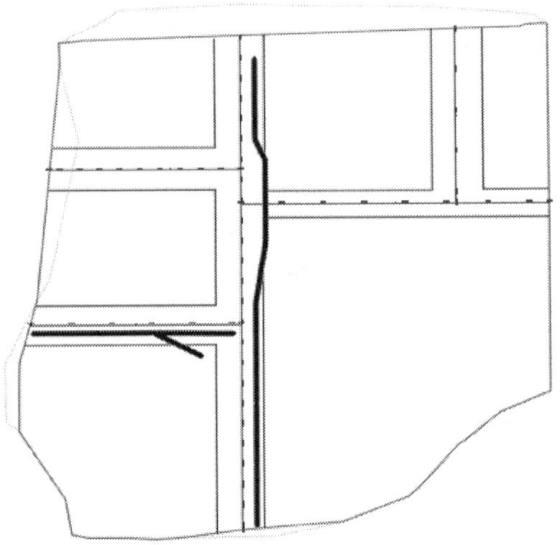

Figure 11-17 *The sanitary sewer network created along the alignments*

Opening the File

1. Open the *c11_c3d_2024_tut.zip* file that you downloaded from *https://www.cadsofttech.com* and saved at *C:\c3d_2024*.

2. Choose **Open** from the **Application Menu**; the **Select File** dialog box is displayed.

3. In the dialog box, browse to the location *C:\c3d_2024\c11_c3d_2024_tut*.

4. Select the file *c11_c3d_2024_tut01.dwg* and then choose the **Open** button to open the file. Ignore any warnings, if displayed. Now, expand the **Surfaces** node in the **Prospector** tab of the **TOOLSPACE** palette; the **EG** and **FG** surfaces will be displayed. In the opened drawing, the **EG** (existing ground) surface is indicated by a yellow border in the drawing

and the **FG** (finished ground) surface is indicated by a red border. The drawing consists of five alignments.

Creating the Parts List

1. Choose the **Create Network Parts List** tool from the **Create Design** panel of the **Home** tab; the **Network Parts List - New Parts List** dialog box is displayed. Choose the **Information** tab.

2. Enter **Sanitary Sewer** in the **Name** edit box of the **Information** tab and then choose the **Pipes** tab.

3. Right-click on **New Parts List** in the **Name** column and choose **Add part family** from the shortcut menu displayed; the **Part Catalog** dialog box is displayed, as shown in Figure 11-18.

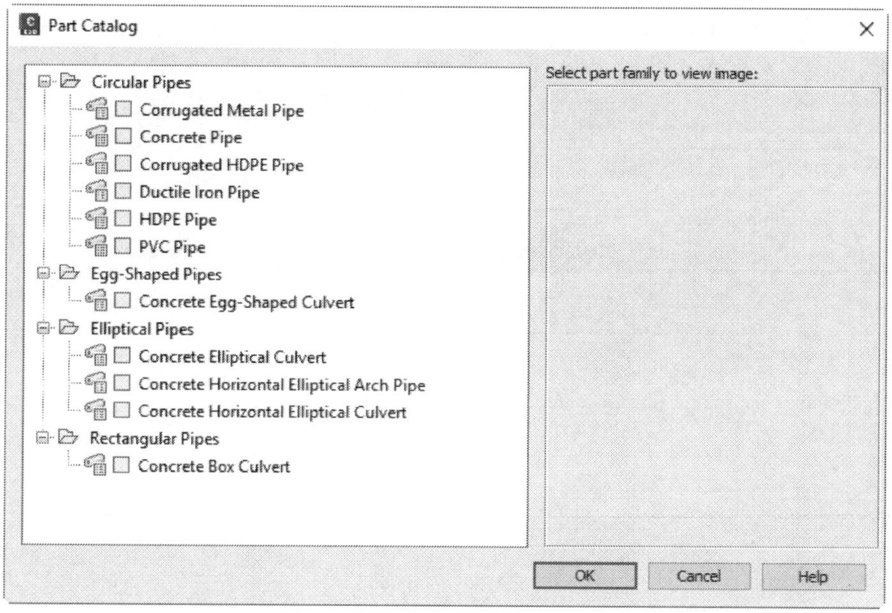

Figure 11-18 *The **Part Catalog** dialog box*

4. Select the **PVC Pipe** check box and then choose the **OK** button; the dialog box is closed and the selected part family is added to the parts list in the **Name** column.

5. Expand the **Sanitary Sewer** sub-node in the **Name** column and right-click on **PVC Pipe**; a shortcut menu is displayed.

6. Choose **Add part size** from the shortcut menu; the **Part Size Creator** dialog box is displayed.

7. Double-click in the **Value** field of the **Inner Pipe Diameter** property and select **6.000000** from the drop-down list, as shown in Figure 11-19. Next, choose the **OK** button; the selected part size is added to the **PVC Pipe** node created in the **Name** column.

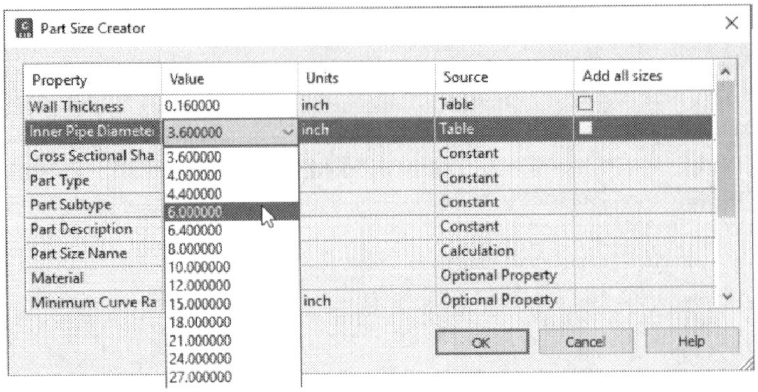

*Figure 11-19 Selecting the **6.000000** value from the **Inner Pipe Diameter** drop-down list*

8. Repeat the procedure followed in steps 6 and 7, and select **10.000000** and **12.000000** as the inner pipe diameters to add two more part sizes. As a result, three pipes of different sizes are added to the **PVC Pipe** node, as shown in Figure 11-20.

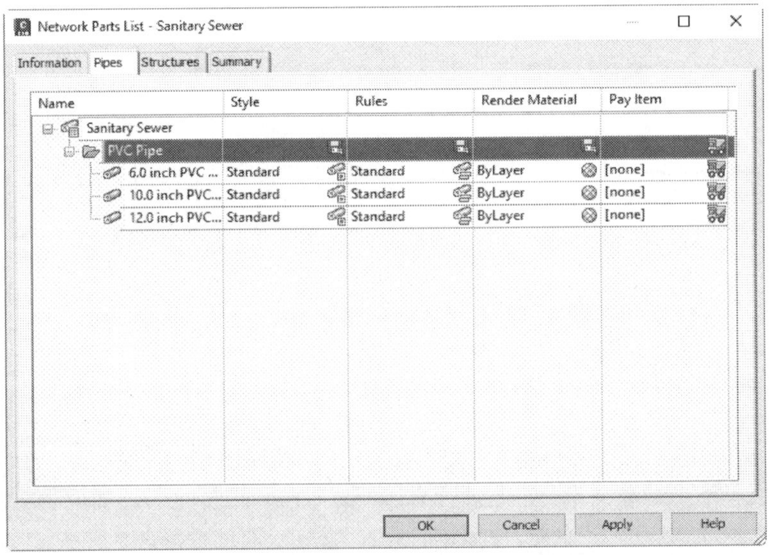

*Figure 11-20 Partial view of three pipes of different sizes added to the **PVC Pipe** node*

9. Now, choose the **Structures** tab in the **Network Parts List - Sanitary Sewer** dialog box and right-click in the dialog box; a shortcut menu is displayed. Choose **Refresh** from the shortcut menu; the item list now displays **Sanitary Sewer** under the **Name** column. Right-click on **Sanitary Sewer** in the **Name** column; a shortcut menu is displayed.

10. Choose the **Add part family** option from the shortcut menu; the **Part Catalog** dialog box is displayed.

11. Select the **Concentric Cylindrical Structure** check box from the **Junction Structures with Frames** category and choose the **OK** button to exit the **Part Catalog** dialog box; the structure part family is added to the part family.

12. Expand the **Sanitary Sewer** and then right-click on **Concentric Cylindrical Structure** in the **Name** column; a shortcut menu is displayed. Choose the **Add part size** option from the shortcut menu; the **Part Size Creator** dialog box is displayed.

13. In this dialog box, ensure that the value of the **Inner Structure Diameter** property is set to **48.000000**. Next, choose the **OK** button to exit. The structure with the specified diameter is added to the part list.

14. Repeat the procedure followed in steps 12 and 13 to add two other structures with the inner structure diameter values as **60.000000** and **72.000000** to the parts list. The structure family and the part sizes are added to the part list, as shown in Figure 11-21.

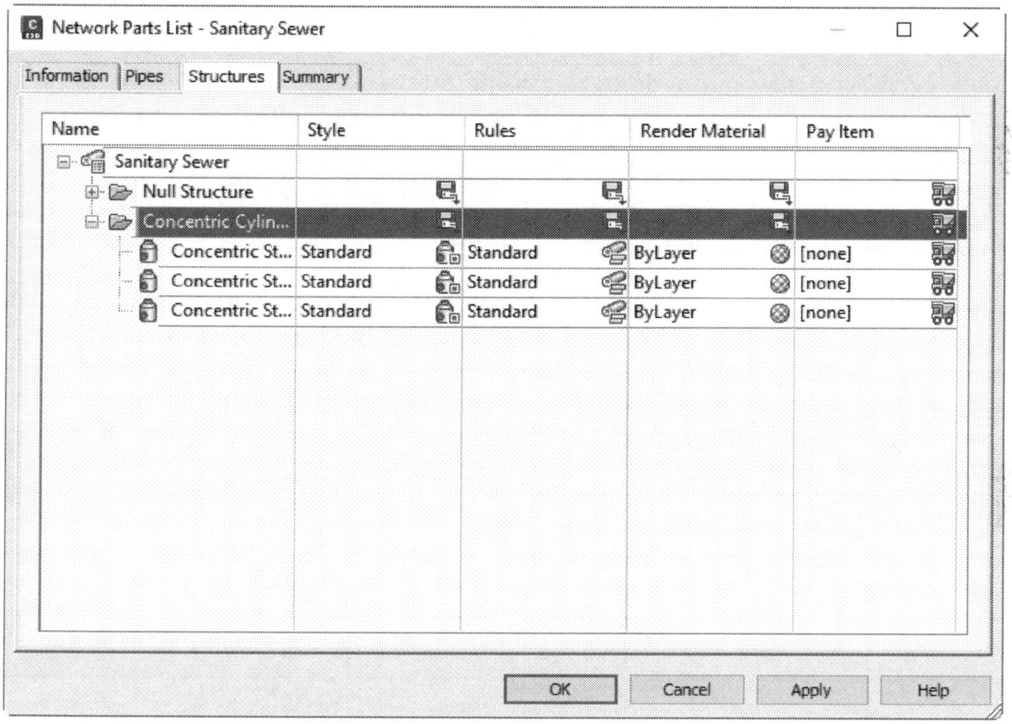

***Figure 11-21** The **Concentric Cylindrical Structure** family and the part sizes added to the **Sanitary Sewer** part list*

15. Choose the **OK** button from the **Network Parts List - Sanitary Sewer** dialog box to close it. Now, you have created the parts list consisting of the required parts. This part list is added to the **Parts Lists** sub-node of the **Pipe Network** in the **Settings** tab and will be used to create the sanitary sewer network.

Creating the Structure Label Style

1. Choose the **Settings** tab from the **TOOLSPACE** palette and expand **Structure > Label Styles** sub-node.

2. Right-click on the **Label Styles** sub-node and choose the **New** option from the shortcut menu displayed; the **Label Style Composer - New Structure Label Style** dialog box is displayed. The **Information** tab is chosen by default in this dialog box.

3. Enter **Structure Style** in the **Name** edit box and then choose the **Layout** tab.

Note that **Structure Text** is the default selected component in the **Component name** drop-down list of this tab.

4. Click in the **Value** field of the **Contents** property and choose the browse button; the **Text Component Editor - Contents** dialog box is displayed. 

5. Select the text in the **Editor** window and delete it.

6. Now, select the **Name** option from the **Properties** drop-down list and choose the button located on the right of this drop-down list; the text **<[Name(CP)]>** is added in the **Text Editor** window. Next, click in the **Editor** window and bring the cursor to the next line.

7. Now, select the **Insertion Rim Elevation** option from the **Properties** drop-down list and set the **Precision** modifier to **0.01**.

8. Again, choose the button located on the right of the **Properties** drop-down list to add the property to the **Text Editor** window, as shown in Figure 11-22.

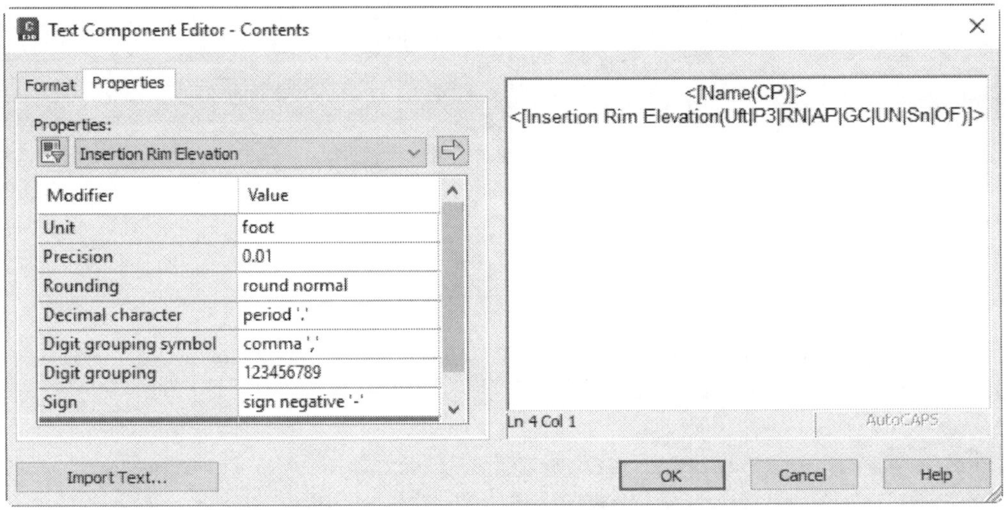

***Figure 11-22** The **Precision** property added to the **Text Editor** window*

9. In the **Editor** window, type **RIM** before the **<Insertion Rim Elevation>** text and align it below the **Name** property.

10. Choose the **OK** button; the **Text Component Editor - Contents** dialog box is closed.

11. In the **Label Style Composer - Structure Style** dialog box, click on the **Create Text component** button; a flyout is displayed. Choose the **Reference Text** option from the flyout, as shown in Figure 11-23; the **Select Type** dialog box is displayed.

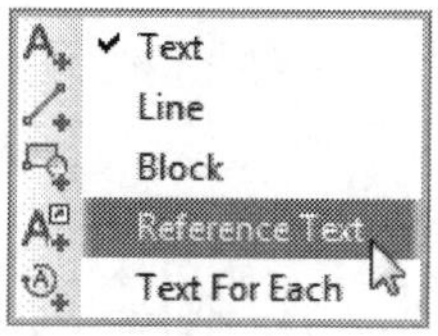

Figure 11-23 *Choosing the* ***Reference Text*** *option from the* ***Create Text Component*** *flyout*

12. Select the **Surface** option from the dialog box and choose the **OK** button; the dialog box is closed.

13. Click in the **Value** field of the **Name** property and enter **FG**.

14. Set the required parameters as given below:

Anchor Component: **Structure Text** Anchor Point: **Bottom Center**

15. Click in the **Value** field of the **Contents** property and choose the browse button; the **Text Component Editor - Contents** dialog box is displayed.

16. Delete the text in the **Editor** window and select the **Surface Elevation** option from the **Properties** drop-down list.

17. Set the **Precision** to **0.01** and choose the arrow button located on the right of the **Properties** drop-down list to add text to the **Editor** window.

18. Enter **FG** before the text in the **Editor** window, as shown in Figure 11-24.

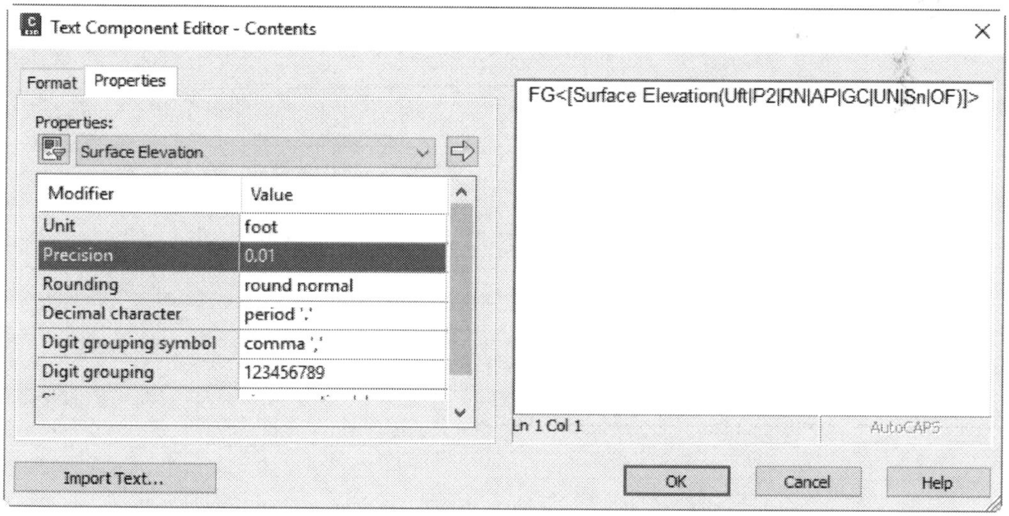

Figure 11-24 *The text entered in the* ***Editor*** *window*

19. Choose the **OK** button to exit the **Text Component Editor - Contents** dialog box and return to the **Label Style Composer - Structure Style** dialog box.

20. Enter **0.3000"** in the **Value** field of the **Text Height** property. You will notice that in the preview area, symbol is displayed in the label as you have not yet selected the surface to be referenced.

Adding Another Component to the Label

1. Choose the down arrow on the right of the **Create Reference Text component** button in the **Layout** tab of the **Label Style Composer - Structure Style** dialog box and choose the **Text For Each** option from the flyout displayed; the **Select Type** dialog box is displayed.

2. Select the **Structure all pipes** option from the list box displayed in the dialog box if it is not selected by default and then choose the **OK** button; the **Select Type** dialog box is closed.

3. In the **Label Style Composer - Structure Style** dialog box, click in the **Value** field of the **Contents** property and then choose the browse button; the **Text Component Editor - Contents** dialog box is displayed.

4. Delete the text in the **Editor** window and select the **Connected Pipe Invert Elevation** option from the **Properties** drop-down list.

5. Set the precision to **0.01** and then add the property to the **Editor window** by using the arrow button located on the right of the **Properties** drop-down list.

6. Add **Invert** before the label text in the **Editor** window and choose the **OK** button to close the **Text Component Editor - Contents** dialog box.

7. In the **Label Style Composer - Structure Style** dialog box, set the parameters as given below:

   Anchor Component: **FG**
   Anchor Point: **Bottom Center**
   Attachment: **Top center**
   Text Height: **0.3000"**

Note
The defined text will be used to display the pipe inverts elevation at the structure.

8. Choose the **OK** button to close the **Label Style Composer - Structure Style** dialog box.

Creating the Sanitary Sewer Pipe Network

1. Choose the **Pipe Network Creation Tools** tool from **Home > Create Design > Pipe Network** drop-down; the **Create Pipe Network** dialog box is displayed.

2. Specify the following values in this dialog box:

   Network name: **Sanitary Sewer Network**
   Network description: **Along the Main Street**
   Network parts list: **Sanitary Sewer**

Surface name: **FG**
Alignment name: **Main Street**
Structure label style: **sewer style**

3. Choose the **OK** button to close the dialog box; the **Network Layout Tools - Sanitary Sewer Network** toolbar is displayed.

4. Select the **Concentric Structure 48 dia 18 frame 24 cone 5 wall 6 floor** option from the **Structure List** drop-down list and then select **10 inch PVC Pipe** from the **Pipe List** drop-down list.

5. Now, invoke the **Pipes and Structures** tool from the drop-down next to the **Pipe List** in the **Network Layout Tools** toolbar; you are prompted to specify the insertion point for the structure.

6. Enter **'SO** in the command line and press ENTER to activate the transparent command for the station offset.

7. Next, select the **Main Street** alignment; you are prompted to specify a station along the alignment.

8. Enter **50** in the command line and press ENTER; you are prompted to specify the offset distance from the station.

9. Enter **-36** in the command line and press ENTER.

10. Follow the prompts in the command line and add a structure at station **200** and **250** each at an offset distance of **-36**, refer to Figure 11-25. You will notice that the structures are placed at the specified stations of the alignment and a pipe connecting the two structures is drawn automatically.

11. Again, enter **350** in the command line and press ENTER to add a structure at station **3+50.00**. Next, enter **-60** as the offset distance and then press ENTER; the structure is placed outside the white boundary line, as shown in Figure 11-26.

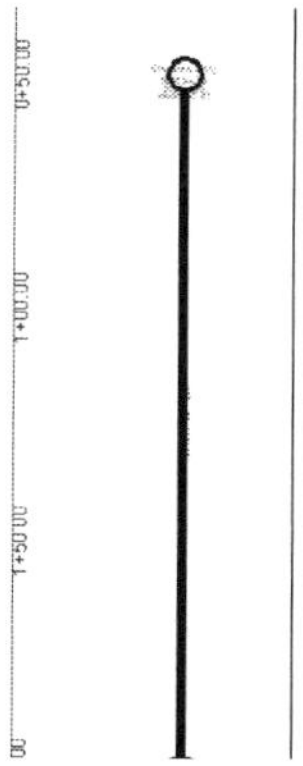

Figure 11-25 *A pipe connecting the structures added at stations*

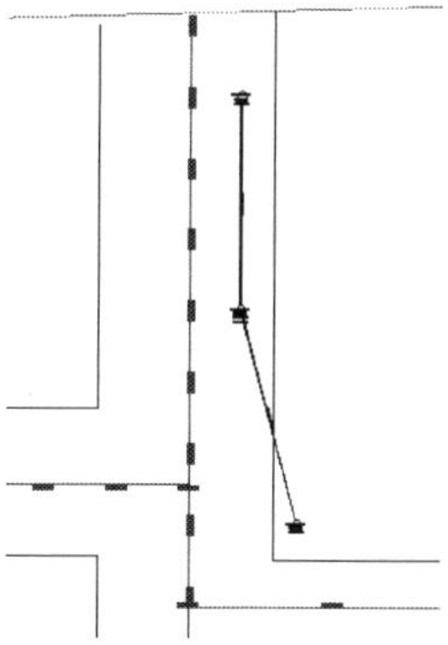

Figure 11-26 *Structures and pipe parts placed outside the boundary*

12. Repeat the procedure given in steps 8 and 9, and add a structure at station **500** at an offset distance of **-60**, at station **650** at an offset distance of **-36**, and at stations **9+00.00**, **10+50.00**, and **12+00.00** each at an offset distance of **-36**.

13. Press ENTER twice to exit the command; the sanitary sewer network along the **Main Street** alignment is created and displayed, as shown in Figure 11-27.

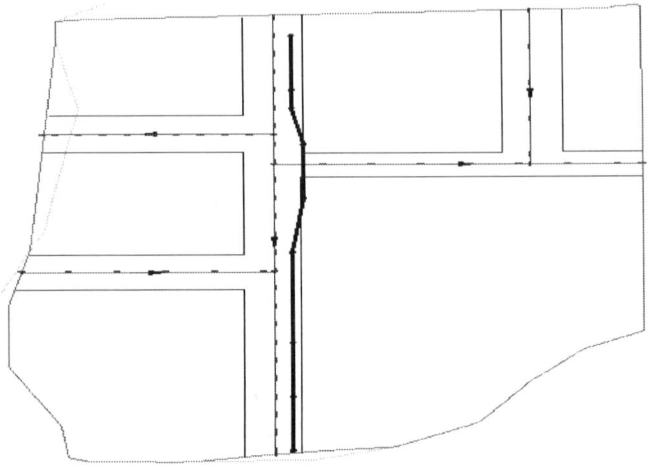

Figure 11-27 *The sanitary sewer network created along the* ***Main Street*** *alignment*

14. Now, select the **Concentric Structure 60 dia 18 frame 24 cone 5 wall 6 floor** option from the **Structure List**. Next, select the **6 inch PVC Pipe** option from the **Pipe List** drop-down list in the **Network Layout Tools** toolbar.

Note

If the ***Network Layout Tools*** *toolbar is not displayed, select any part of the network from the drawing and right-click. Choose* ***Edit Network*** *from the shortcut menu displayed to enable the toolbar.*

15. Invoke the **Draw Pipes and Structures** tool from the **Network Layout Tools** toolbar and click near the first station of the **MG Road** alignment (second alignment on the left) to add a structure, as shown in Figure 11-28.

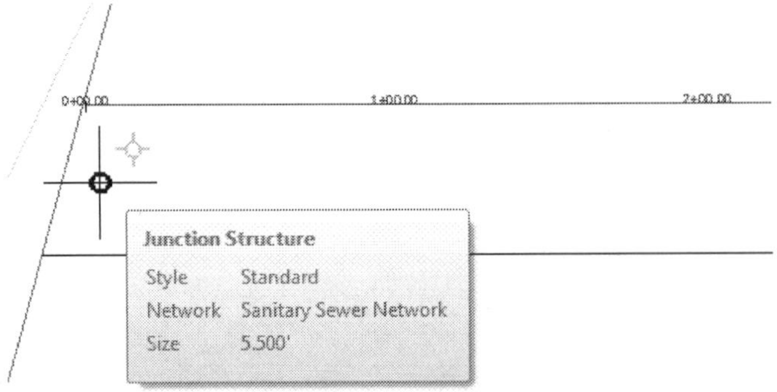

Figure 11-28 *Adding a structure near the first station of the* ***MG Road*** *alignment*

16. Next, click again between the stations **1+00.00** and **2+00.00** to place another structure.

17. Now, select the **10 inch PVC Pipe** option from the **Pipe List** drop-down list and click in the drawing area to add structures near stations **3+00.00** and **5+00.00**.

18. Press ENTER to exit the command.

19. Select **Concentric Structure 60 dia 18 frame 24 cone 5 wall 6 floor** from the **Structure List** drop-down list and **12 inch PVC Pipe** from the **Pipe List** drop-down list in the **Network Layout Tools** toolbar.

20. Choose the **Structures Only** tool from the drop-down next to the **Pipe List** in the **Network Layout Tools** toolbar, as shown in Figure 11-29.

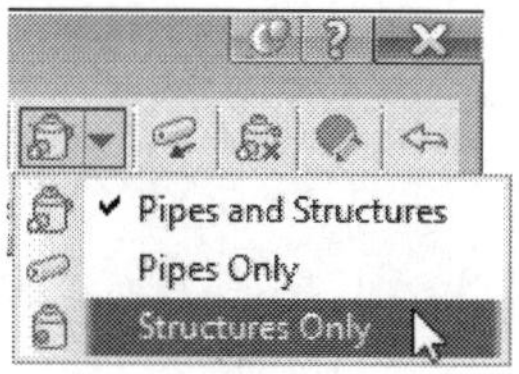

Figure 11-29 *Choosing the **Structures Only** option from the flyout*

21. Now, add a structure outside the alignment right-of-way, as shown in Figure 11-30.

22. Choose **Pipes Only** from the flyout.

23. Move the cursor near the structure added in step 21, and click when the orange colored circular glyph is displayed, as shown in Figure 11-31. This glyph indicates the insertion or the connection point to connect the pipe.

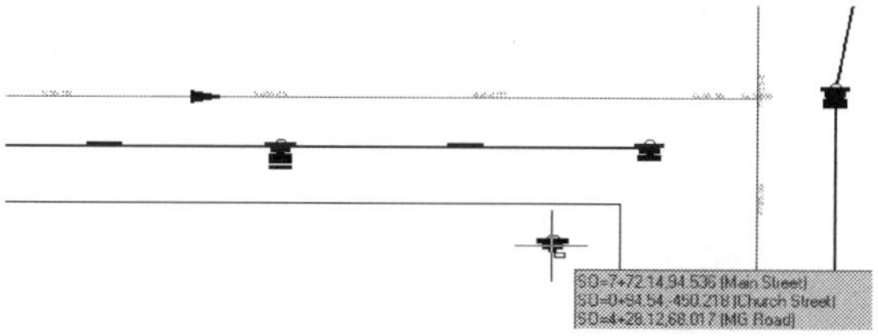

Figure 11-30 *Adding the structure outside the right-of-way*

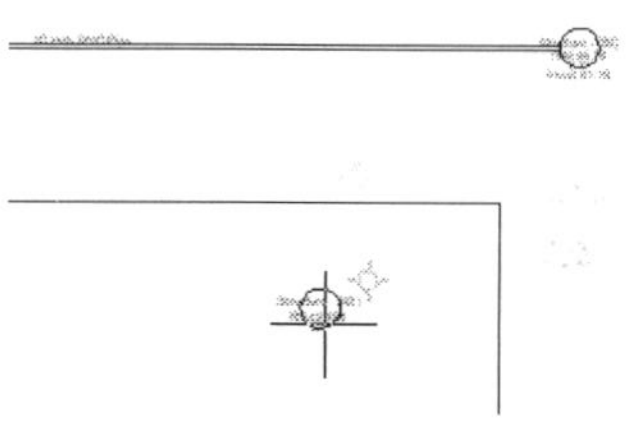

Figure 11-31 *The orange colored glyph displayed at the insertion point*

24. Now, move the cursor to the structure added near station **3+00.00**; the orange glyph is displayed. Next, click in the drawing; the pipe connecting the two structures is created, as shown in Figure 11-32.

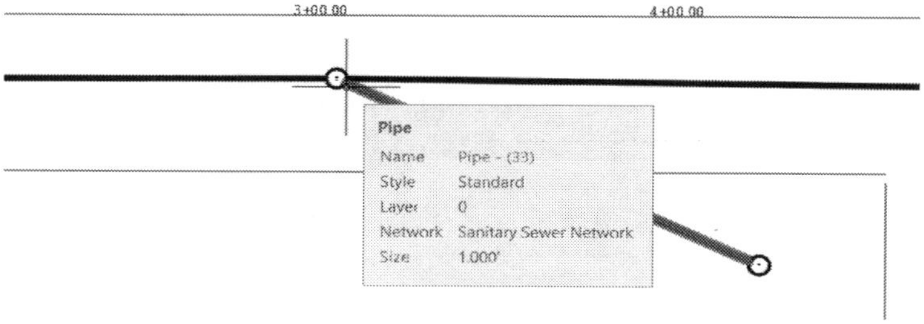

Figure 11-32 *The pipe created between two structures*

25. Press ENTER to terminate the command.

Referencing the Label Text

1. Zoom in the center of the structure of any of the structure labels and you will notice that the symbol is displayed in the structure label as mentioned earlier. To get rid of this symbol, you need to reference the required surface to the label text.

2. Select any of the structure labels from the drawing and right-click; a shortcut menu is displayed.

3. Choose **Label Properties** from the shortcut menu to display the **PROPERTIES** window.

4. Click on the default value **none** of the **sewer style::FG::surface** property under the **Reference Text Objects** head; a drop-down list is displayed.

5. Select **FG** from the drop-down list, as shown in Figure 11-33. Close the **PROPERTIES** window.

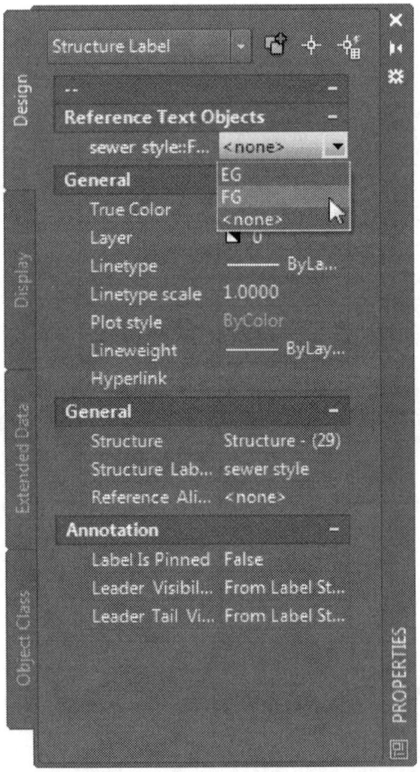

***Figure 11-33** Selecting the **FG** option from the drop-down list of the **sewer style** property*

6. Zoom in the selected structures again; you will notice that the symbol is replaced with the elevation of the **FG** surface, refer to Figures 11-34 and 11-35.

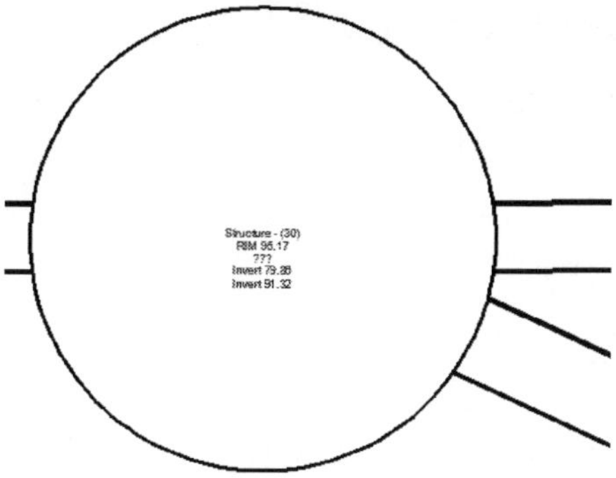

Figure 11-34 The structure label without the referenced object

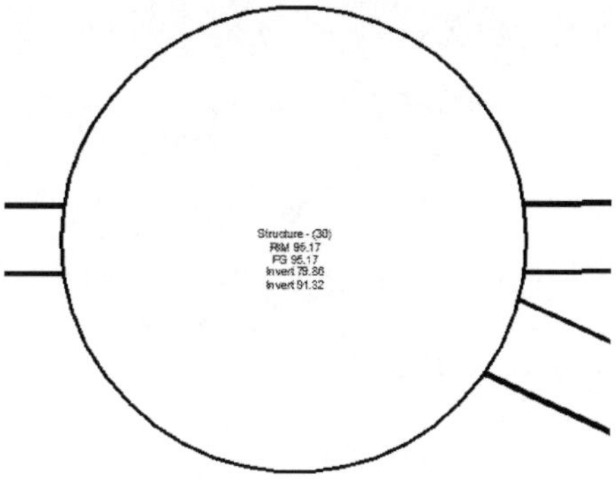

Figure 11-35 The structure label after referencing the surface object with the label

7. Zoom in to any other structure; you will notice the symbol. To reference the value in all the structures in one go, right-click on any structure label; a shortcut menu is displayed. Choose the **Select Similar** option from the shortcut menu displayed; all the labels for the structures will be selected. Now, right-click and select the **FG** surface as the reference object for the label text, as explained in steps 3 to 5.

8. Zoom in to check the structure labels and view the elevation of the **FG** surface.

Saving the File

1. Choose **Save As** from the **Application Menu**; the **Save Drawing As** dialog box is displayed.

2. In this dialog box, browse to the following location:

   C:\c3d_2024\c11_c3d_2024_tut

3. Save the file as *c11_tut01a*.

EXERCISE

1. Log on to *www.cadsofttech.com* and browse to *Books > Civil/GIS Books > AutoCAD Civil 3D > AutoCAD Civil 3D 2024 for Novices*. Next, select *c11_c3d_2024_exer.zip* file from the **Exercise Files** drop-down list. Choose the **Download** button to download the data file.

2. Now, save and extract the downloaded folder to the following location:

   C:\c3d_2024

Note

*While opening the exercise files, the **PANORAMA** window may appear with an error message. Close this window to proceed further.*

Exercise 1

Use the *c11_c3d_2024_tut02.dwg* file and create a water pipeline parts list and water pipeline network, as shown in Figure 11-36, using the **Network Layout Tools** toolbar. Also, run the interference check. **(Expected time: 30 min)**

Use the following parameters:

Parts List name:	**Water Pipeline**
Pipe Parts:	**Concrete Pipe**
Part Size:	**36.000000 inch inner pipe diameter**
Structure Part:	**No**
Network Name:	**Water Pipe Network** (**Create Pipe Network dialog box**)
Network parts List:	**Water Pipeline**
Surface name:	**Finished Surface**
Alignment name:	**Rosewood street**
Toggle Upslope/DownSlope:	**UpSlope** (**Network Layout Tools** toolbar)
Network Tool:	**Draw Pipes Only tool**
Starting station of the pipe network:	**6+48**
Station Offset:	**18**

Click after an interval of 100 at every station till the station 3+00. For example, the stations can be at 6+00, 5+00, 4+00, and 3+00 and then at 2+50, 2+00, 100, and 0+00 all at an offset distance of 18, refer to Figure 11-36. Also, run the interference check between the sanitary sewer and the water pipeline network using the 3D Proximity check criteria. Use the distance of 18 inch as the tolerance.

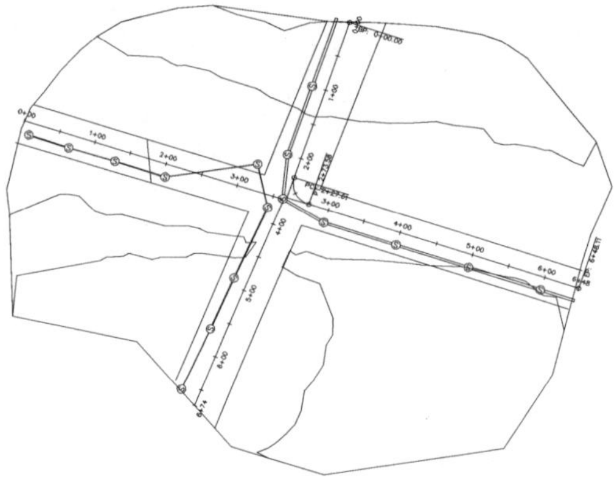

Figure 11-36 *The water pipeline network created along the* ***Rosewood street*** *alignment*

Chapter 12

Pressure Networks

Learning Objectives

After completing this chapter, you will be able to:

- *Define pressure network components*
- *Create the pressure network in a plan layout*
- *Draw and refine the pressure network in a profile view*
- *Create a pressure network from objects*
- *Edit and access pressure network properties*
- *Add labels and tables to pressure network objects*
- *Validate a pressure network*

PRESSURE NETWORK

Pressure network is a network consisting of pressure pipes, fittings, and appurtenances in which pressure energy is used to deliver a liquid or a gaseous substance from the storage source to the destination. In AutoCAD Civil 3D, a set of tools are used to design pressure networks. To design a pressure network, you need to define a list of network components called parts list. After defining the parts list, you will create the pressure network in horizontal plane and refine the network design in vertical plane. The procedure to define network components (parts list) is discussed next.

Defining Network Components

To design a pressure network, you first need to create a parts list. A parts list contains a list of components that will be used in the network and is derived from the part catalog. A part catalog enlists the components of the pressure network. To create a parts list, first you need to set a network catalog and then create parts list.

Creating a Pressure Network

Ribbon:	Home > Create Design > Pipe Network drop-down > Pressure Network Creation Tools
Command:	CREATEPRESSURENETWORK

Civil 3D provides a set of tools to design a pressure network. You will design a pressure network in a horizontal layout and then refine the networks elevation information in profile view.

To create a pressure network, choose the **Pressure Network Creation Tools** tool from the **Home > Create Design > Pipe Network** drop-down; the **Create Pressure Pipe Network** dialog box is displayed, as shown in Figure 12-1. In this dialog box, the **Network name** textbox displays the name of the pressure network. To specify a name, choose the **Click to edit the name template** button located next to the text box; the **Name Template** dialog box will be displayed. Enter the desired name in the **Name** edit box of this dialog box and choose the **OK** button to exit. Optionally, you can include the information about the network in the **Network description** text box.

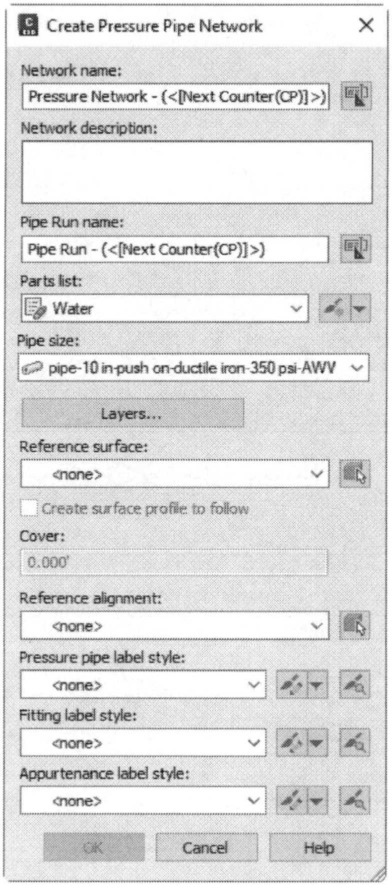

***Figure 12-1** The **Create Pressure Pipe Network** dialog box*

Creating Pressure Networks from Objects

Ribbon:	Home > Create Design > Pipe Network drop-down > Create Pressure Network from Object
Command:	CREATEPRESSURENETWORKFROMOBJ

In AutoCAD Civil 3D, existing Civil 3D objects such as lines, polylines, arcs, and feature lines can be converted into a network of connected pressure pipes and bends. AutoCAD Civil 3D converts these entities to a pressure pipe network and then inserts structures at the end points or vertices of these entities. To create a pressure pipe network from an object, choose the **Create Pressure Network from Object** tool from the **Create Design** panel; you will be prompted to select an object or an Xref. Select the existing entity in the drawing; arrows will be displayed on the selected entity showing the direction of flow. Press ENTER to accept the default direction or enter **R** in the command line and then press ENTER; the direction of flow will be reversed. Again, press ENTER; the **Create Pressure Pipe Network from Object** dialog box will be displayed, as shown in Figure 12-2.

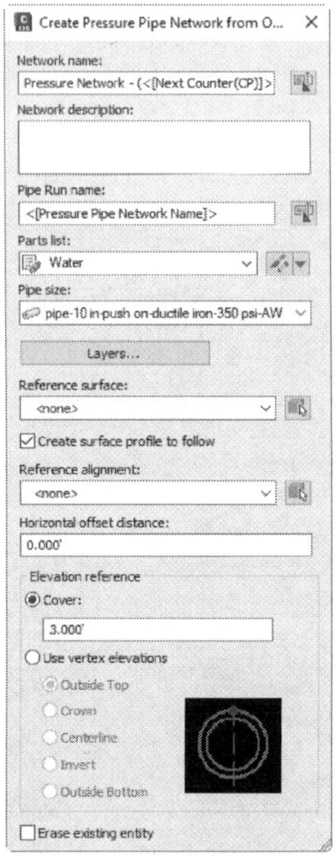

Figure 12-2 The ***Create Pressure Pipe Network from Object*** *dialog box*

PRESSURE NETWORK PROPERTIES

To access and edit the pressure pipe network properties such as network name, labels, layers, profile, and section view settings, select any part of the network in the drawing and right-click; a shortcut menu will be displayed. Choose the **Pressure Network Properties** option from the shortcut menu displayed; the **Pressure Pipe Network Properties - <network name>** dialog box will be displayed, refer to Figure 12-3. The **Information** tab of this dialog box displays the name of the selected pipe network. Modify the name in the **Name** edit box, if required.

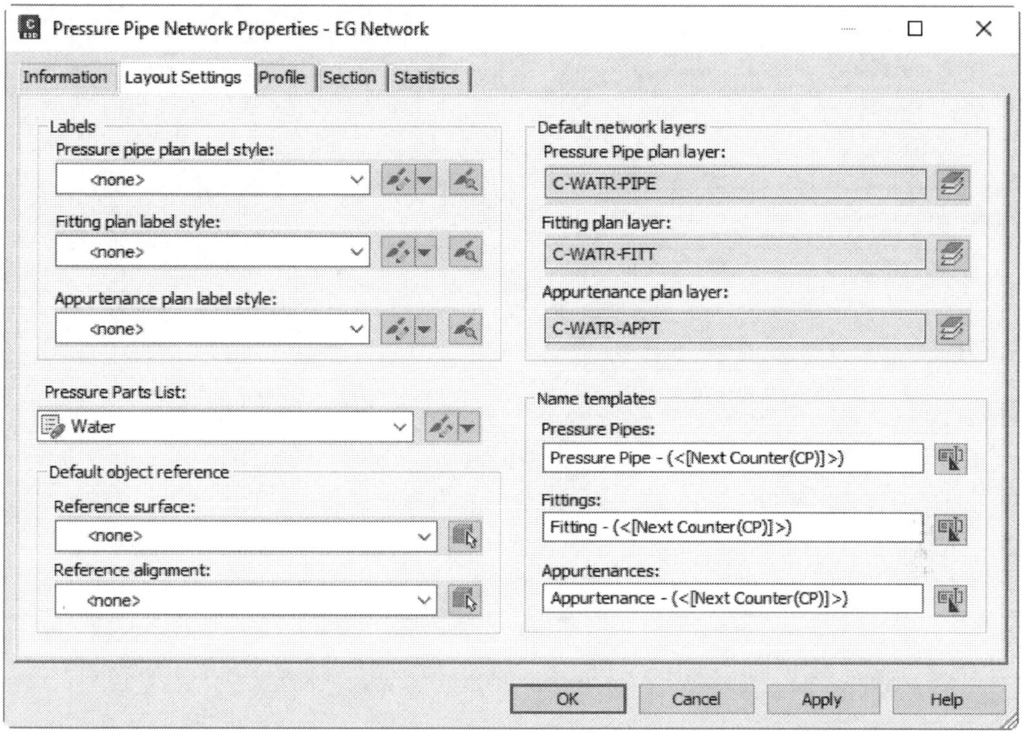

Figure 12-3 *The* ***Pressure Pipe Network Properties - EG network*** *dialog box*

Pressure Pipe Properties

To view and edit the pressure pipe properties, select the required pressure pipe part from the drawing and right-click; a shortcut menu will be displayed. Next, choose the **Pressure Pipe Properties** option from this menu; the **Pressure Pipe Properties - <pipe name>** dialog box will be displayed, as shown in Figure 12-4.

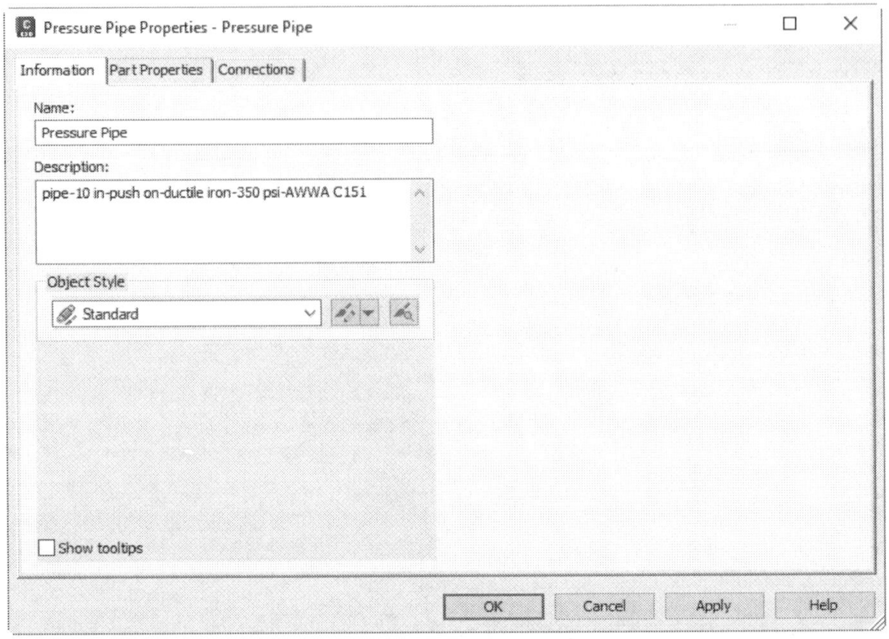

Figure 12-4 The ***Pressure Pipe Properties - <pipe name>*** *dialog box*

Fittings and Appurtenances Properties

To view and edit the fittings or appurtenances properties, select the required fitting or appurtenance and right click; a shortcut menu will be displayed. Depending upon the part selected, choose the **Fitting Properties** or the **Appurtenance Properties** from this menu; the **Fitting Properties - <fitting name>** or **Appurtenance Properties - <appurtenance name>** dialog box will be displayed. The options in these dialog boxes are same as in the **Pressure Pipe Properties - <pipe name>** dialog box which have already been explained in the **Pressure Pipe Properties** section of this chapter.

EDITING PRESSURE NETWORK

Civil 3D provides you with different methods for editing the pressure networks in both the plan and profile views. You can use the **Pressure Network Plan Layout: <Network Name>** tab or the **Pressure Network Profile Layout: <Network Name>** tab for editing the network in the plan and profile views, respectively. You can also use the **TOOLSPACE** item view and the grip editing methods to edit the pressure network in both the plan and profile views as required. These editing methods are discussed next.

Using the Pressure Network Plan Layout or Profile Layout Tab

Ribbon:	Pressure Networks: <Network Name> > Modify > Edit Network
Command:	EDITPRESSURENETWORK

To edit pressure network in plan view, select it; the **Pressure Networks: <Network Name>** contextual tab is displayed. Select the **Plan Layout Tools** option from the **Edit Network** drop-down list in the **Modify** panel of this tab; the **Pressure Network Plan Layout - <Network**

Name> tab will be displayed. You can choose various tools from this tab to add, change, or delete the network parts. You can also modify the associated surface and the alignment. All the options and commands in this tab have been explained in detail in the **Creating Pressure Networks > Designing a Pressure Network in Plan Layout** section of this chapter.

Editing Pressure Network Objects Using the TOOLSPACE Item View

You can also edit the pressure network objects such as pressure pipes, fittings, and appurtenances using the **TOOLSPACE** item view. To do so, choose the **Prospector** tab in the **TOOLSPACE** palette and expand **Pressure Networks > Networks** sub-node. Next, expand the required network and select any of the pressure network objects. On doing so, the data for the selected object will be displayed in the **Item View** list area. Now, click in the desired column cells to edit the required values.

Grip Editing of the Pressure Pipe Network

AutoCAD Civil 3D gives you the flexibility of using grips to edit the pressure networks. Each time you select a pressure network object, the control grips will be displayed. You can use these grips to move or resize the objects. For example, select the required pressure pipe from the plan view; square grips will be displayed at the center of the pipe, as shown in Figure 12-5.

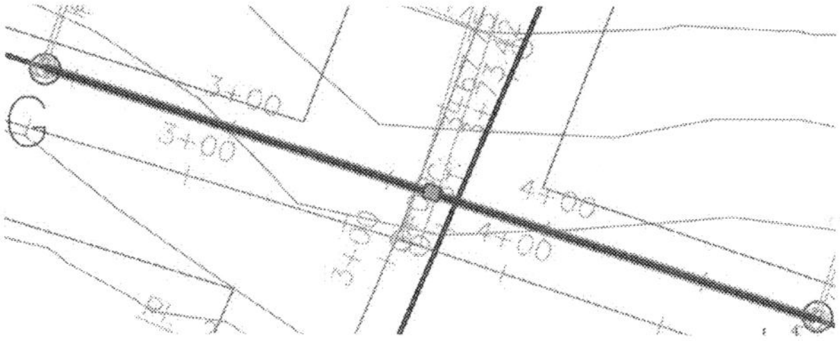

Figure 12-5 *Grips displayed on selecting the pipe object before changing the location*

ADDING PRESSURE NETWORK LABELS

Ribbon:	Annotate > Labels & Tables > Add Labels drop-down > Pressure Network > Add Pressure Networks Labels
Command:	ADDPRESSURENETWORKLABELS

You can label the pressure network or some parts of the network using the in-built label styles. The label styles can be created and managed in the **Settings** tab of the **TOOLSPACE** palette. You can access the pressure pipes, fittings, and appurtenance labels styles in the **Pressure Pipe**, **Fitting**, and **Appurtenance** nodes, respectively. The labels can be added while creating the network or can also be added after the pressure network is created. To add the labels while creating the network, select the required label style from the **Create Pressure Pipe Network** dialog box.

To add the pressure network labels, after the pressure network is created, choose **Pressure Network > Add Pressure Network Labels** from the **Add Labels** drop-down, as shown in Figure 12-6. On doing so, the **Add Labels** dialog box will be displayed, as shown in Figure 12-7.

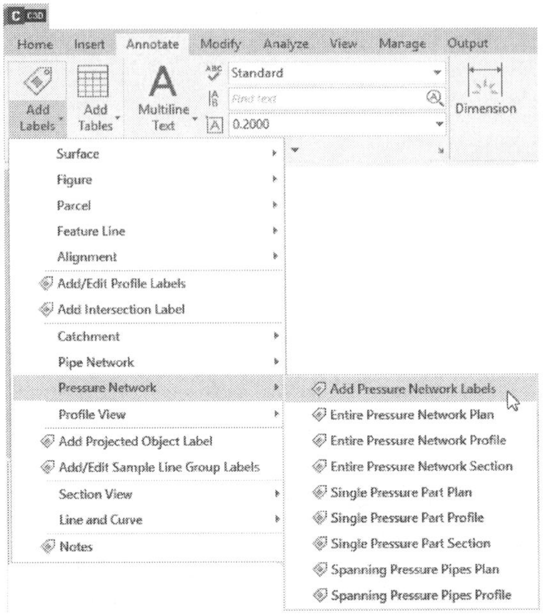

Figure 12-6 *Choosing the* ***Add Pipe Network Labels*** *tool from the flyout*

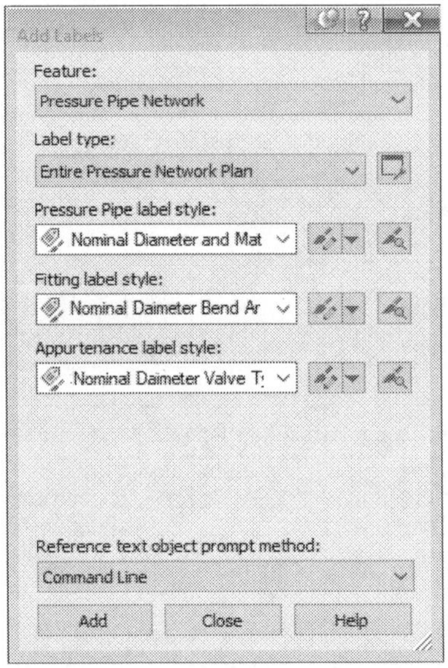

Figure 12-7 *The* ***Add Labels*** *dialog box*

ADDING PRESSURE NETWORK TABLES

To view the pressure network objects data in a tabular form, you can add the tables in the drawing. These tables help you to view the consolidated information about the objects that are used to draw the pressure network.

Adding a Pressure Pipe Table

Ribbon:	Annotate > Labels & Tables > Add Tables drop-down > Pressure Network > Add Pressure Pipe
Command:	ADDPRESSUREPIPETABLE

To add the table for structures, choose **Pressure Network > Add Pressure Pipe** from the **Add Tables** drop-down; the **Pressure Pipe Table Creation** dialog box will be displayed, as shown in Figure 12-8. In this dialog box, select the required option from the **Table style** drop-down list. You can also create a new style for the pressure pipe table. To do so, choose the down arrow button on the right of the **Table style** drop-down list; a flyout will be displayed. Choose **Create New** option from the flyout; the **Table Style - New Pressure Pipe Table Style** dialog box will be displayed. Specify the required options in this dialog box and choose the **OK** button; the dialog box is closed.

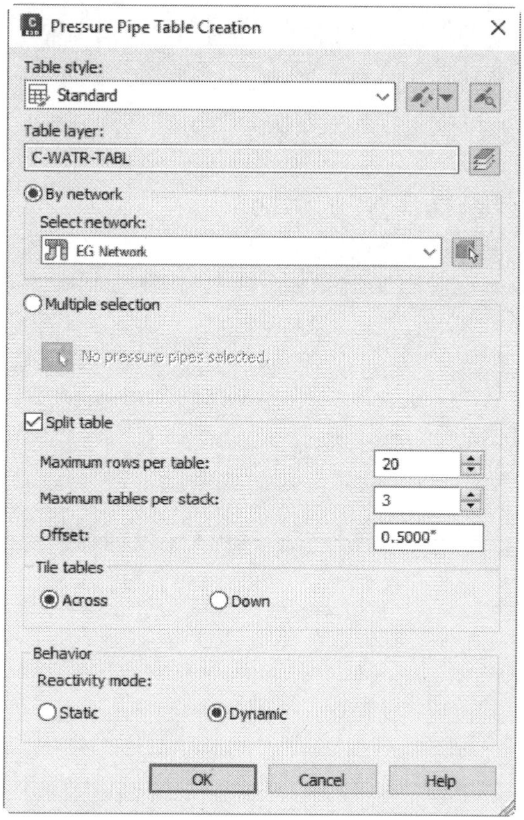

Figure 12-8 *The* ***Pressure Pipe Table Creation*** *dialog box*

Adding a Fitting Table

Ribbon:	Annotate > Labels & Tables > Add Tables drop-down > Pressure Network > Add Fitting
Command:	ADDFIITINGTABLE

To add a table for the fittings, choose **Pressure Network** > **Add Fitting** from the **Add Tables** drop-down; the **Fitting Table Creation** dialog box will be displayed. The options in this dialog box are same as discussed for the **Pressure Pipe Table Creation** dialog box. After you have specified the settings and selected the required network or the fitting objects, choose the **OK** button; the **Fitting Table Creation** dialog box will be closed and you are prompted to specify the upper left corner of the table. Click in the drawing to specify the table corner; the **Fitting Table** will be added to the drawing displaying the fitting details, as shown in Figure 12-9. Like the **Pressure Pipe Table**, you can configure the settings for the **Fitting Table** in the table style and display the required information about the fitting object in the table.

Fitting Table		
Fitting Name	Horizontal Angle	Vertical Angle
Fitting - (1)	11.2393	0.4907
Fitting - (3)	14.7551	17.0427
Fitting - (4)	9.7227	5.7080

Figure 12-9 *The* ***Fitting Table*** *displaying the fitting details*

Adding an Appurtenance Table

Ribbon: Annotate > Labels & Tables > Add Tables drop-down > Pressure Network > Add Appurtenance
Command: ADDAPPURTTABLE

To add a table for appurtenances, choose the **Pressure Network** > **Add Appurtenance** from the **Add Tables** drop-down; the **Appurtenance Table Creation** dialog box will be displayed. The options in this dialog box are same as discussed for the **Pressure Pipe Table Creation** dialog box and by using these options, you can create a table for the appurtenances.

VALIDATING A PRESSURE NETWORK

In AutoCAD Civil 3D, you can validate the designed pressure network created by you by running a design check and a depth check on the network. Performing these checks helps in verifying the pressure network for various design parameters such as deflection, diameter, open connections, radius of curvature, and so on. You can also check whether the designed pressure network is within the allowed minimum and maximum depth.

After running the design and the depth checks, Civil 3D highlights the area where the pressure network violates the specified settings and parameters by placing the warning symbols into the drawing. When you hover the cursor over a warning symbol, a tooltip will be displayed which describes the violation.

Note

Both the design check and the depth check are run separately on a pressure network because the design check can only be run on multiple branches of the pressure network, whereas the depth check can be run on one branch at a time.

Performing Design Check

As mentioned above, by running the design check on a pressure network, you can check the network for parameters such as deflection, diameter, open connections, and radius of curvature. To run the design check, select the required pressure network graphically; the **Pressure Network Plan Layout:<network name>** contextual tab will be displayed. Next, choose **Design Check** from the **Analyze** panel of this tab; the **Run Design Check** dialog box will be displayed, as shown in Figure 12-10. In this dialog box, from the **Select parameter to check** area, select check boxes corresponding to the parameters for which you need to validate the pressure network and then choose the **OK** button; the **Run Design Check** dialog box will be closed and the warning symbols will be displayed automatically at the location where pressure network violates the specified parameters.

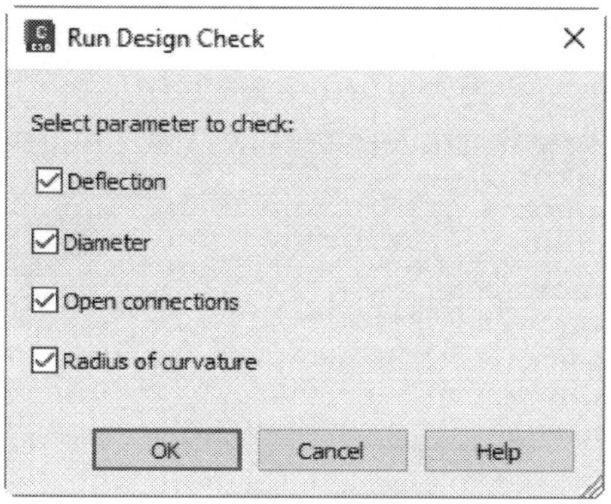

***Figure 12-10** The **Run Design Check** dialog box*

Performing Depth Check

In a pressure pipe network, pipe elevations are determined according to the specified depth, below a terrain from the ground level. Therefore, running a depth check ensures that the specified minimum cover value is satisfied along the full length of the pressure pipe and also validates that both the minimum and the maximum cover values are not violated at any length of the pressure pipe network. To run the depth check, select the required pressure network pipe; the **Pressure Network Plan Layout:<Network Name>** contextual tab will be displayed. Next, choose **Depth Check** from the **Analyze** panel of this tab; you will be prompted to select a path along the pressure network in the plan or profile view. Select the required pressure pipe and press ENTER; the **Run Depth Check** dialog box will be displayed. In this dialog box, under the **Select parameter to check** area, specify the minimum and maximum cover of depth in the **Minimum depth of cover** and **Maximum depth of cover** edit boxes, respectively. Note that, you can enter the values in these edit boxes only after selecting their respective check box, refer to Figure 12-11. Next, choose the **OK** button; the **Run Depth Check** dialog box will be closed. Now, if the selected pressure pipe violates the value of any of the specified parameters then a warning symbol will be displayed in the drawing. However the cursor over the warning symbol; a tooltip showing the violation information is displayed.

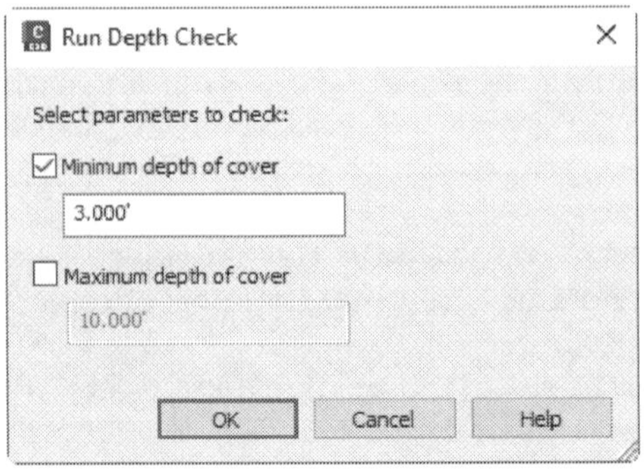

***Figure 12-11** The **Run Depth Check** dialog box*

TUTORIAL

Before starting the tutorial, you need to download and save the tutorial files on your computer. To do so, follow the steps given below:

1. Log on to *www.cadsofttech.com* and browse to *Books > Civil/GIS Books > AutoCAD Civil 3D > AutoCADCivil 3D 2024 for Novices*. Next, select *c12_c3d_2024_tut.zip* file from the **Tutorial Files** drop-down list. Choose the **Download** button to download the data file.

2. Now, save and extract the downloaded folder to the following location:

   C:\c3d_2024

Note

*While opening the tutorial file, the **PANORAMA** window may appear with an error message. Close this window to proceed further.*

Tutorial 1 Pressure Network - II

In this tutorial, you will create a pressure network from a polyline object, as shown in Figure 12-12 and run the design and the depth checks on the created pressure network. Also, you will add labels to the created pressure network. **(Expected time: 30 min)**

***Figure 12-12** The Pressure network created from the polyline object*

The following steps are required to complete this tutorial:

a. Open the file.
b. Create the pressure network from the polyline object in the drawing.
c. Run the design and depth checks.

d. Add labels to the network.
e. Save the file.

Opening the File

1. Open the *c12_c3d_2024_tut.zip* file that you downloaded from *https://www.cadsofttech.com* and saved at *C:\c3d_2024*.

2. Choose **Open > Drawing** from the **Application Menu**; the **Select File** dialog box is displayed.

3. In this dialog box, browse to location *C:\c3d_2024\c12_c3d_2024_tut* where you have saved the file.

4. Select the file *c12_c3d_2024_tut01.dwg* and then choose the **Open** button to open it. This drawing file consists of a surface, a profile view, and an alignment which you will use to create the pressure network.

Creating the Pressure Network

1. Choose the **Create Pressure Network from Object** tool from **Home > Create Design >Pipe Network** drop-down; you are prompted to select an object.

2. Zoom in to the drawing and select the polyline object from the drawing displayed below the alignment; you are prompted to select the flow direction.

3. Press ENTER to accept the default direction; the **Create Pressure Pipe Network from Object** dialog box is displayed.

4. Enter the following details in the dialog box:

Network name:	**Water Main**
Network parts list:	**Water**
Size and material:	**pipe-12 in- push on- ductile iron- 350 psi- AWWA C151**
Surface name:	**EG surface**
Alignment name:	**EG Road**
Depth of cover:	**3.00'**

5. Also, select the **Erase existing entity** check box and choose the **OK** button; the **Create Pressure Pipe Network from Object** dialog box is closed and the pressure network is created from the selected entity.

6. Select any pressure pipe entity from the network and right-click; a shortcut menu is displayed.

7. Choose the **Edit Pressure Pipe Style** option from the shortcut menu; the **Pressure Pipe Style- Standard** dialog box is displayed. In this dialog box, choose the **Display** tab.

8. In the **Component display** area of this tab, select the **Color** cell corresponding to the **Outside Pipe Walls** parameter; the **Select Color** dialog box is displayed. Choose the red color from this dialog box and choose the **OK** button; the **Select Color** dialog box is closed; the red color is assigned to the **Outside Pipe Walls** parameter.

9. Similarly, assign red color to the **Inside Pipe Walls** and **Pipe End Line** parameters.

10. Next, choose the **OK** button in the **Pressure Pipe Style- Standard** dialog box; the dialog box is closed.

Notice that the color of the pressure pipe in the drawing has changed and it is displayed in red color.

Running the Design and the Depth Checks

1. Select all the pressure pipes of network; the **Pressure Networks: Water Main** contextual tab is displayed.

2. Choose the **Design Check** option from the **Analyze** panel of this tab; the **Run Design Check** dialog box is displayed.

3. In this dialog box, ensure that the **Diameter**, **Open connections**, and **Radius of curvature** check boxes in the **Select parameter to check** area are selected and then choose the **OK** button; the **Run Design Check** dialog box is closed and a message is displayed in the command line that design check is complete.

   Two violations are detected and marked as yellow colored warning symbols in the drawing.

4. Hover the cursor over the warning symbol; a tooltip describing the violation is displayed, as shown in Figure 12-13.

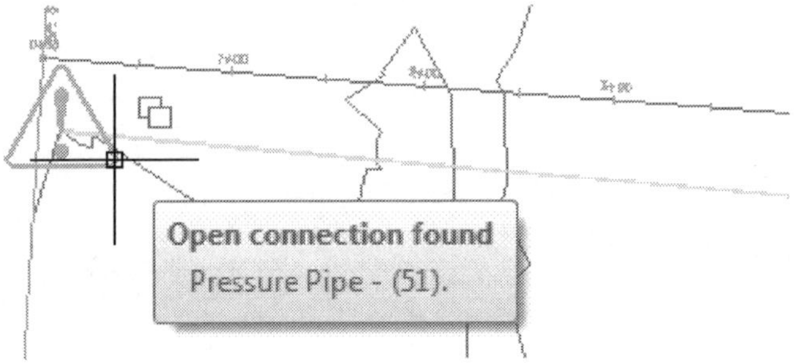

Figure 12-13 *Yellow colored warning symbol with the tooltip showing the violation*

5. Now, select the **Depth Check** option from the **Analyze** panel of the **Pressure Networks: WaterMain** contextual tab; you are prompted to select a path along the pressure network in the plan or profile view.

6. Next, select the left most pipe segment in the network; you are prompted to select the next point on the path.

7. Select the right most pipe segment; all the pipes in the pressure network are selected.

8. Next, press ENTER; the **Run Depth Check** dialog box is displayed.

9. In this dialog box, in the **Select parameters to check** area, select the check box corresponding to **Minimum depth of cover** if not selected by default and then enter the value **2.00'** in the edit box.

10. Choose the **OK** button; the **Run Depth Check** dialog box is closed and a message is displayed in the command line that two warnings symbols marking the violations are displayed in the drawing, as shown in Figure 12-14.

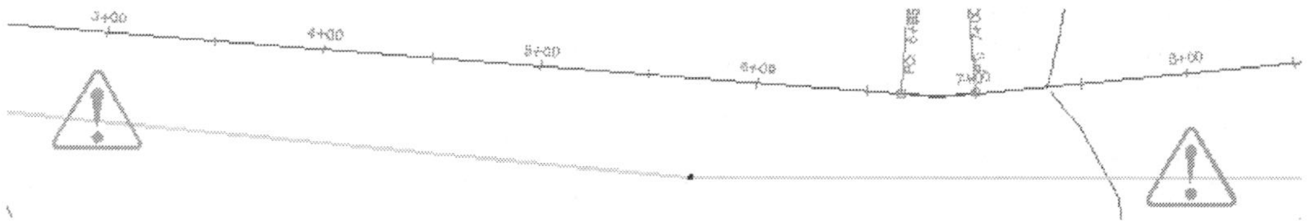

Figure 12-14 *Warning symbols showing the violation for depth*

Adding Pressure Network Labels

1. Choose the **Pressure Network** option from **Annotate > Labels & Tables > Add Labels** drop-down; a flyout is displayed. From this flyout, choose the **Add Pressure Network Labels** option; the **Add labels** dialog box is displayed.

2. Ensure that the **Entire Pressure Network Plan** option is selected in the **Label type** drop-down list.

3. Next, select the **Standard** option from the **Pressure Pipe label style** and **Fitting label style** drop-down lists.

4. Now, choose the **Add** button; you are prompted to select a part in the pressure network. Select any part of the **Water Main** pressure network; the pressure pipes and the fittings of the pressure network are labeled. Next, choose the **Close** button to close the **Add Labels** dialog box.

The added label shows the pressure pipe and fitting name or number and the name of the layer on which the pressure network is created, refer to Figure 12-15.

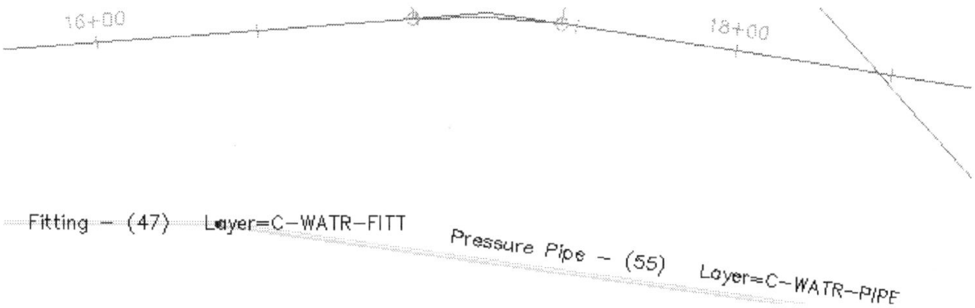

Figure 12-15 *The pressure pipe and fitting labels added to the* ***Water Main*** *pressure network*

Saving the File

1. Choose **Save As** from the **Application Menu**; the **Save Drawing As** dialog box is displayed.

2. In this dialog box, browse to the following location:

   C:\c3d_2024\c12_c3d_2024.tut

3. Save the file as *c12_tut01a*.

EXERCISE

1. Log on to *www.cadsofttech.com* and browse to *Books > Civil/GIS Books > AutoCAD Civil 3D > AutoCAD Civil 3D 2024 for Novices*. Next, select *c12_c3d_2024_exer.zip* file from the **Exercise Files** drop-down list. Choose the **Download** button to download the data file.

2. Now, save and extract the downloaded folder to the following location:

   C:\c3d_2024

Note

*While opening the exercise file, the **PANORAMA** window may appear with an error message. Close this window to proceed further.*

Exercise 1

Open the *c12_c3d_2024_ex01.dwg* file and create a pressure network from the feature line in the drawing, refer to Figure 12-16. Also, add labels to the pressure network.

(Expected time: 30min)

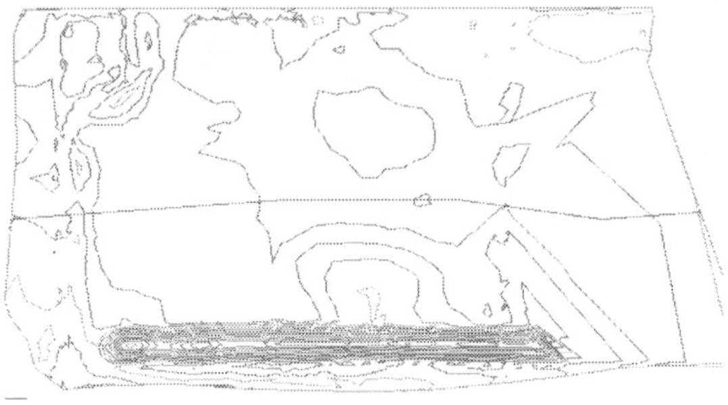

Figure 12-16 *The feature line from which the pressure network will be created*

Network name:	**Water Main**
Network parts list:	**Water**
Size and material:	**pipe-12 in- push on- ductile iron- 350 psi- AWWA C151**
Surface name:	**Ground surface**
Depth of cover:	**3.00'**

Chapter 13

Working with Plan Production Tools and Data Shortcuts

Learning Objectives

After completing this chapter, you will be able to:

- *Understand the plan production tools*
- *Create data shortcuts*
- *Create data references*
- *Promote and synchronize references*

PLAN PRODUCTION TOOLS

AutoCAD Civil 3D provides you with comprehensive tools that help you create construction documents and sheet sets containing the layout of plan and profile views. These tools are known as plan production tools. Sheet sets display the alignment sections and the profile views.

In AutoCAD Civil 3D, drawing templates, namely *Civil 3D (Imperial) Plan over Plan.dwt* , *Civil 3D (Metric) Plan over Plan.dwt*, *Civil 3D (Imperial) Profile over Profile.dwt*, and *Civil 3D (Metric) Profile over Profile.dwt*, have been added to create plan/plan and profile/profile sheets. In addition, new plan production tools are introduced that enable you to create construction documents more quickly by allowing you to create plan(s) only, profile(s) only sheets.

DATA SHARING

Any Civil Engineering project requires a team of engineers to work simultaneously for effective implementation and management of the project. This would require effective data management and data sharing tools. AutoCAD Civil 3D provides you with such effective tools. These tools help the team members to access the project data simultaneously and work independently on the project. These tools are discussed next in detail.

Data Shortcut

The data shortcut is a method that allows you to create external references for different Civil 3D objects. These references can be inserted into one or more drawings and can be worked upon by multiple users simultaneously. The references or data shortcuts are XML files representing a shortcut to Civil 3D objects and contain information about the objects for identifying them in the source drawings. Using the data shortcut, you can divide a project or model into separate drawings. These drawings are called source drawings and contain the objects that you can share. Prior to AutoCAD Civil 3D, data shortcuts can only be created for alignments, surfaces, profiles, view frame groups, and pipe networks. Now, you can create data shortcuts for corridors also. When a data shortcut is created for corridors, the data shortcuts for alignments and profiles related to the corridors are automatically created. Additionally, you can save the geometry of the referenced surface into the current drawing. The objects in the source drawings can be used in other drawings provided you create a data reference for it.

Note

*The data shortcuts are created and managed in the **Master View** of the **TOOLSPACE** palette.*

Data Reference

A data reference is a link to use data shortcut objects in a new drawing. The referenced objects provide you with the flexibility to apply a temporary object style, annotation, and perform various analyses with them. The properties of these referenced objects can also be viewed in the drawing.

Data reference is a two step process, first is to create a data shortcut and second is to create a data reference.

TUTORIAL

Before starting the tutorial, you need to download and save the tutorial files on your computer. To do so, follow the steps given below:

1. Log on to *www.cadsofttech.com* and browse to *Books > Civil/GIS Books > AutoCAD Civil 3D > AutoCAD Civil 3D 2024 for Novices*. Next, select *c13_c3d_2024_tut.zip* file from the **Tutorial Files** drop-down list. Choose the **Download** button to download the data file.

2. Now, save and extract the downloaded folder to the following location:

   C:\c3d_2024

Note

While opening the tutorial file, the ***PANORAMA*** *window may appear with an error message. Close this window to proceed further.*

Tutorial 1 View Frames and Sheets

In this tutorial, you will create view frames and sheets using plan production tools, refer to Figure 13-1. **(Expected time: 30 min)**

The following steps are required to complete this tutorial:

a. Open the file.
b. Create view frames.
c. Create sheets.
d. Save the file.

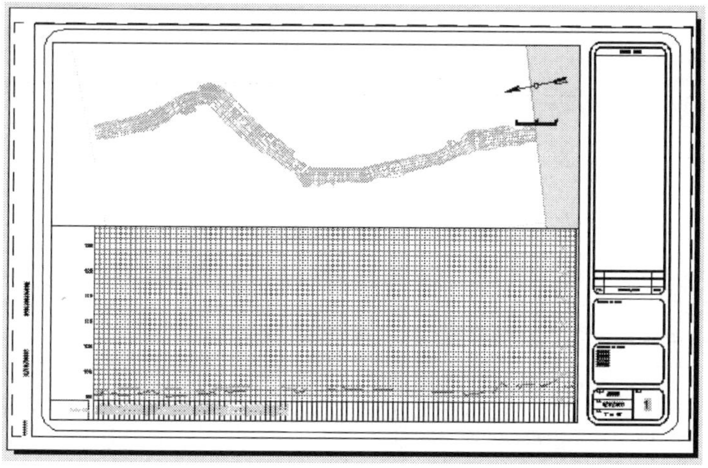

Figure 13-1 *The sheet with plan and profile views*

Opening the File

1. Open the *c13_c3d_2024_tut.zip* file that you downloaded from *https://www.cadsofttech.com,* as explained in Chapter 2.

2. Choose **Open > Drawing** from the **Application Menu**; the **Select File** dialog box is displayed.

3. In this dialog box, browse to the location *C:\c3d_2024\c13_c3d_2024_tut* and select the file *c13_c3d_2024_tut01.dwg*.

4. Choose the **Open** button to open the file. Ignore the error message if displayed in the **PANORAMA** window.

Creating View Frames

1. Choose the **Create View Frames** tool from the **Plan Production** panel of the **Output** tab; the **Create View Frames** wizard with the **Alignment** page is displayed.

2. In the **Alignment** page of this wizard, select the **Stephen Road** option from the drop-down list in the **Alignment** area if it is not selected by default.

3. Select the **Automatic** radio button in the **Station Range** area of the **Alignment** page of the wizard.

4. Choose the **Next** button; the **Sheets** page of the **Create View Frames** wizard is displayed.

5. In the **Sheet Settings** area of the **Sheets** page, select the **Plan and Profile** radio button. Next, choose the browse button displayed on the right of the **Template for Plan and Profile sheet** edit box; the **Select Layout as Sheet Template** dialog box is displayed, as shown in Figure 13-2.

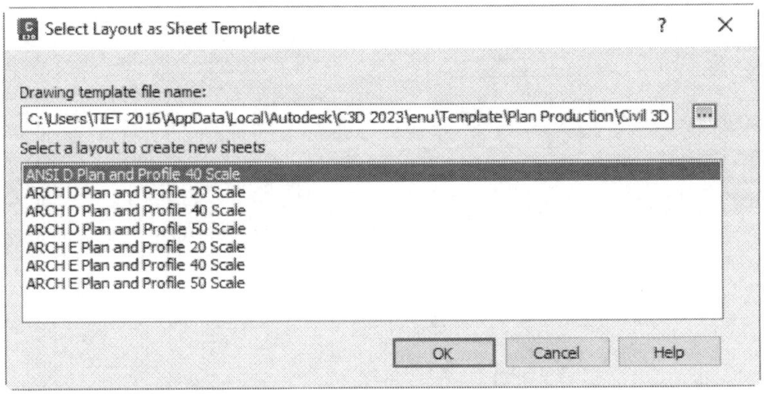

***Figure 13-2** The **Select Layout as Sheet Template** dialog box*

6. In this dialog box, choose the browse button on the right of the **Drawing template file name** edit box; the **Select Layout as Sheet Template** dialog box is displayed.

7. In this dialog box, browse to the location *C:\Users\<username>\AppData\Local\Autodesk\C3D 2024\enu\Template\Plan Production* and then select the *Civil 3D (Imperial) Plan and Profile.dwt* template file if it is not selected by default.

8. Choose the **Open** button to return to the **Select Layout as Sheet Template** dialog box. In this dialog box, ensure that the **ANSI D Plan and Profile 40 Scale** option is selected in the **Select a layout to create new sheets** list box. Next, choose the **OK** button; the dialog box is closed.

9. In the **View Frame Placement** area of the **Sheets** page, make sure that the **Along alignment** radio button is selected. Choose the **Next** button; the **View Frame Group** page of the **Create View Frames** wizard is displayed.

10. In this page, enter **Stephen group** in the **Name** edit box of the **View Frame Group** area.

11. Accept all the default values in the **View Frame** area and choose the **Next** button; the **Match Lines** page of the wizard is displayed.

12. Accept the default values in this page and then choose the **Next** button; the **Profile Views** page of the **Create View Frames** wizard is displayed.

13. Retain the default settings in this page and choose the **Create View Frames** button from the wizard; the view frames are created, as shown in Figure 13-3.

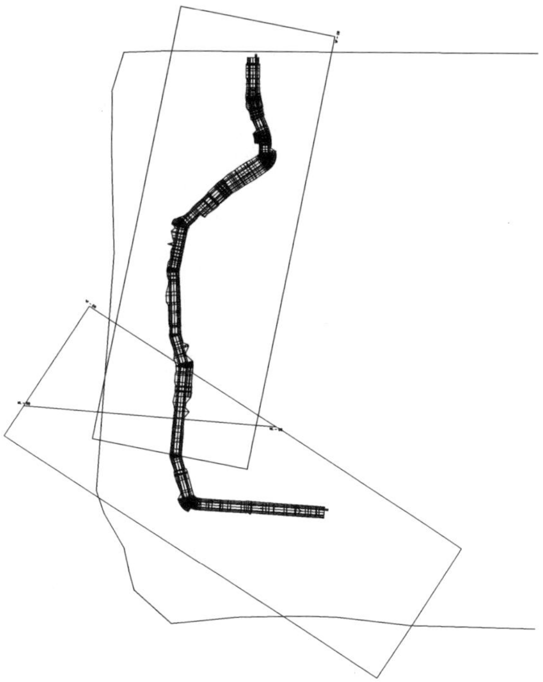

*Figure 13-3 The **Stephen group** view frames*

Creating Sheets

1. Choose the **Create Sheets** tool from the **Plan Production** panel of the **Output** tab; the **View Frame Group and Layouts** page of the **Create Sheets** wizard is displayed.

2. In the **Layout Creation** area of this page, enter **Stephen sheet** in the **Layout name** edit box.

3. In the **Choose the north arrow block to align in layouts** drop-down list, select the **North** option.

4. Retain all other default settings in the **View Frame Group and Layouts** page and choose the **Next** button; the **Sheet Set** page of the **Create Sheets** wizard is displayed.

5. In the **Sheet Set** area, ensure that the **New sheet set** radio button is selected. Next, choose the Browse button on the right of the **Sheet set storage location** edit box; the **Browse for Sheet Set Folder** dialog box is displayed.

6. Browse to the location *C:\Users\<user name>\Documents\Autodesk* in this dialog box and select the folder *My Civil 3D Tutorial Data*.

Note
If the My Civil 3D Tutorial Data folder is not available at C:\Users\<user name>\Documents\Autodesk, create a new folder and rename it as My Civil 3D Tutorial Data.

7. Choose the **Open** button; the **Browse for Sheet Set Folder** dialog box is closed and the **Sheet Set** page of the **Create Sheets** wizard is displayed.

8. In the **Sheets** area of the **Sheet Set** page of the wizard, choose the Browse button next to the **Sheet files storage location** edit box; the **Browse for Folder** dialog box is displayed.

9. Navigate to the location *C:\Users\<user name>\Documents\Autodesk\My Civil 3D Tutorial Data* and choose the **Open** button to return to the **Sheet Set** page.

10. Enter **Stephen sheet file** in the **Sheet file name** edit box and choose the **Next** button; the **Profile Views** page is displayed.

11. Retain the default settings in the **Profile Views** page and choose the **Next** button to display the **Data References** page.

12. Ensure that the **Alignments** check box is selected. Select the **Surfaces** check box to draw the surface in the sheet.

13. Choose the **Create Sheets** button in the wizard; the **AutoCAD Civil 3D 2024** message box is displayed, as shown in Figure 13-4.

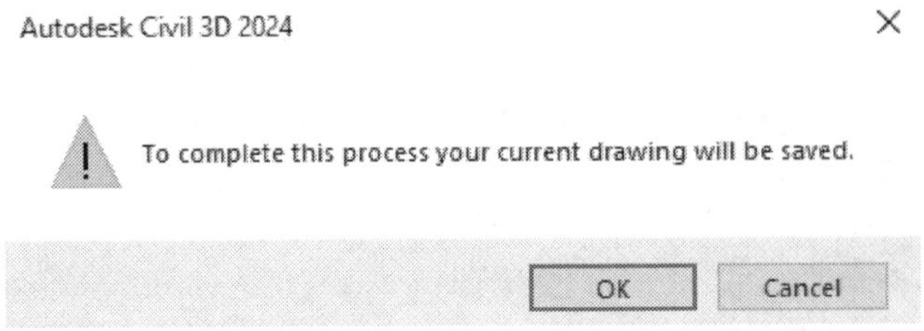

***Figure 13-4** The **AutoCAD Civil 3D 2024** message box*

14. Choose the **OK** button; the message box is closed and you are prompted to specify the origin for the profile view.

15. Click in the drawing area; the **Create Sheets Progress Dialog** window is displayed showing the progress of the sheets creation. After the sheet creation process is completed, the **SHEET SET MANAGER** palette and the **PANORAMA** window are displayed. Close the **PANORAMA** window.

16. Right-click on the first sheet **1-Stephen sheet** in the **SHEET SET MANAGER** palette and then choose the **Open** option from the shortcut menu displayed, as shown in Figure 13-5. On doing so, the sheet is displayed in a new drawing, as shown in Figure 13-6. In addition, the **PANORAMA** window is displayed. Ignore the message and close the **PANORAMA** window.

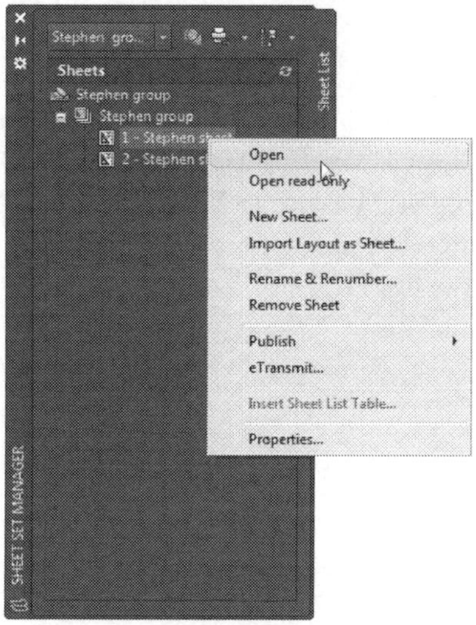

Figure 13-5 *Choosing the **Open** option from the shortcut menu*

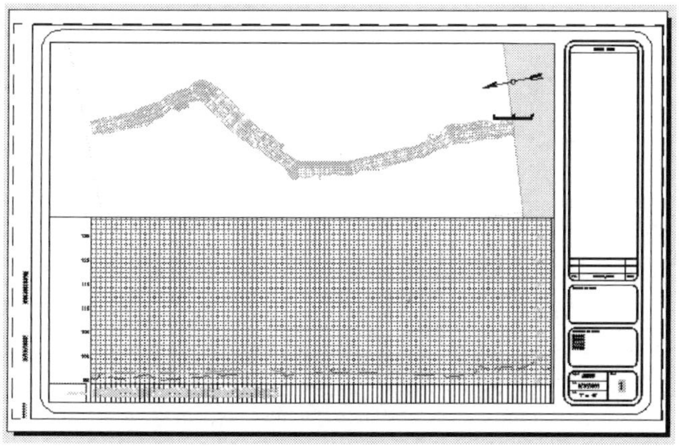

Figure 13-6 *The sheet displayed in a new drawing*

Saving the File

1. Choose **Save As** from the **Application Menu**; the **Save Drawing As** dialog box is displayed.

2. In this dialog box, browse to the following location:

   C:\c3d_2024\c13_c3d_2024_tut

3. In the **File name** edit box, enter **c13_tut01a**.

4. Choose the **Save** button; the file is saved with the name *c13_tut01a.dwg* at the specified location.

5. Choose the **Close** button from the **Application Menu**; the file is closed.

EXERCISE

1. Log on to *www.cadsofttech.com* and browse to *Books > Civil/GIS Books >AutoCAD Civil 3D > AutoCAD Civil 3D 2024 for Novices*. Next, select *c13_c3d_2024_exer.zip* file from the **Exercise Files** drop-down list. Choose the **Download** button to download the data file.

2. Now, save and extract the downloaded folder to the following location:

   C:\c3d_2024

Note

*While opening the exercise file, the **PANORAMA** window may appear with an error message. Close this window to proceed further.*

Exercise 1

Open the *c13_c3d_2024_ex01.dwg* drawing file and create the view frames shown in Figure 13-7. Also, create the sheets using the view frames. **(Expected time: 30 min)**

Figure 13-7 *Surface with view frames*

Index

Q

S

T

U

V

W

This page is intentionally left blank

Other Publications by CADSoft Technologies

The following is the list of some of the publications by CADSoft Technologies. Please visit *www.cadsofttech.com* for the complete listing.

Revit Architecture for Novices Textbooks
- Revit Architecture 2024 for Novices: Learn By Doing
- Revit Architecture 2023 for Novices: Learn By Doing
- Revit Architecture 2020 for Novices: Learn By Doing
- Revit Architecture 2019 for Novices: Learn By Doing

Revit Structure for Novices Textbooks
- Revit Structure 2023 for Novices: Learn By Doing
- Revit Structure 2020 for Novices: Learn By Doing

Revit MEP for Novices Textbooks
- Revit MEP 2020 for Novices: Learn By Doing
- Revit MEP 2019 for Novices: Learn By Doing

Autodesk 3dx Max for Novices Textbooks
- Autodesk 3ds Max 2020 for Novices: Learn By Doing
- Autodesk 3ds Max 2019 for Novices: Learn By Doing

Autodesk Maya for Novices Textbook
- Autodesk Maya 2019 for Novices: Learn By Doing

Pixologic ZBrush for Novices Textbook
- Pixologic ZBrush 2018 for Novices: Learn By Doing

MAXON CINEMA 4D for Novices Textbook
- MAXON CINEMA 4D R20 Studio for Novices: Learn By Doing

AutoCAD for Novices Textbooks
- AutoCAD 2020 for Novices: Learn By Doing
- AutoCAD 2019 for Novices: Learn By Doing
- AutoCAD 2018 for Novices: Learn By Doing

Inventor Professional for Novices Textbook
- Inventor Professional 2020 for Novices: Learn By Doing

Siemens NX for Novices Textbook
- Siemens NX 2019 for Novices: (Continous Release) Learn By Doing

Creo Parametric for Novices Textbook

- Creo Parametric 6.0 for Novices: Learn By Doing
- Creo Parametric 5.0 for Novices: Learn By Doing

SOLIDWORKS for Novices Textbook

- SOLIDWORKS 2018 for Novices: Learn By Doing